Thesis Zhang
29 - 31, 04

Geometry: An Intuitive Approach

Margaret Wiscamb Hutchinson

University of St. Thomas

Charles E. Merrill Publishing Company

A Bell & Howell Company

Columbus, Ohio

Merrill Mathematics Series

Erwin Kleinfeld, Editor

International Standard Book Number: 0-675-09427-5

Library of Congress Catalog Card Number: 79-171538

1 2 3 4 5 6 7 8 9 10—79 78 77 76 75 74 73 72

Printed in the United States of America

To Eve and Scott

Preface

Geometry is a study that has several aspects. From one approach geometry is a logical deductive system. This was the viewpoint the ancient Greeks took and it is, more or less, the way geometry is taught in high schools today. At the other extreme we might take an applied approach to geometry. This was the viewpoint of the ancient Egyptians and this approach is used every day by carpenters, surveyors, architects, draftsmen, navigators and many others. This practical approach might be called the geometry of measure since it involves careful measurements and, in certain cases, meticulous diagrams. The Egyptians accomplished marvels of engineering and their surveyors were exceedingly skillful. They constructed right angles with great accuracy; however they did not know, nor did they care, *why* a triangle measuring 3, 4 and 5 units on a side would have a right angle opposite the longest side.

In this book we will attempt to steer a middle course between strictly theoretic geometry on the one hand and applied geometry on the other. Our approach will be informal and intuitive. We will state some axioms and use these to prove a number of theorems; however some useful theorems will be stated without proof. We will make liberal use of diagrams and sketches to aid our intuition. Some theorems will only be suggested and the student led to discover them for himself. One of the aims of this book is to develop the student's geometric intuition, and this approach might be called "intuitive geometry."

The book has many problem exercises. While some of them are routine drill problems, many are problems which are planned to lead the student to explore topics which are not specifically discussed in the text. For example, in the chapter on congruence of triangles there are a number of problems which explore congruence of rectangles.

If there is one idea that the student should gain from this book it is that Euclidean geometry is not some God-given universal truth but strictly a creation of man. In fact, man has created not just one, but many geometries—all equally valid. This is discussed in Chapter 2 in a section on mathematical models and again in Chapter 6 in the section on Non-Euclidean geometry.

This book was written for those who in their teaching will be concerned with elementary geometric concepts. It may follow a course in number systems; however, it is in no way dependent on such a course and may even preceed it. Little or no mathematical background on the part of the student is assumed. The recommendations of CUPM (The Committee on the Undergraduate Program in Mathematics) have been followed rather closely.

This book has been used in a one term three-hour course. If it follows a course on number systems, the material in Chapter 1 on sets may be omitted or reviewed briefly. Chapter 5 on geometric constructions could very easily be assigned as work outside of class. The instructor then might reasonably expect to cover Chapters 1, 2, 3, 4, 6, 7, 8, and 9. For students with a strong background a review of Chapters 2, 3, 4, and 6 could be followed by a more intensive study of Chapters 7, 8, 9, 10, and 11.

I would like to express my warmest thanks and appreciation to my husband, Professor J. D. Hutchinson of the University of Houston, who read the manuscript and made many helpful suggestions; to Miss Frances Kopczynski for her skillful and conscientious work in typing the manuscript, to Miss Pat Bush who helped with the sketches and finally to my students in Math 138 who acted as "guinea pigs" while this was being written.

Contents

Contents

1: Set Theory and Logic

1. Sets and Subsets

While geometry is one of the most ancient branches of mathematics, set theory is one of the most modern. Created barely a hundred years ago by Georg Cantor, the ideas of set theory have revolutionized mathematics. Since we will use these simple but basic ideas in our study of geometry, let us take time here to review them briefly.

A *set* is simply some collection of objects. We already use the word in this sense when we refer to a set of dishes, or a set of blocks. We will use capital letters A, B, C, . . . to name sets. For example, we might say, "Let S be the set of all triangles." The objects in the set are called *elements* of the set or *members* of the set, and will be denoted by small letters. If an element p belongs to a set S, we write, symbolically, $p \in S$. The symbol " $\in$ " may be read "is an element of," "is a member of," or even "belongs to."

If p is *not* in the set S, then we can express this symbolically $p \notin S$. The slash through the symbol " $\in$ " expresses negation, and " $\notin$ " is read "is not an element of," etc.

For example, if N is the set of counting numbers, (counting numbers are the numbers we count with—1, 2, 3, and so on) then $2 \in N$, (the number 2 belongs to the set N) but $0 \notin N$. (Zero is not in N.)

There are several ways of describing a set. The simplest is to list all the members of the set. Traditionally, the elements are written down separated by commas, and the entire list is enclosed in braces. For example,

$$N = \{1, 2, 3, \ldots\}$$
$$A = \{\text{John, Paul, Mary}\}$$
$$C = \{\text{Austin}\}$$

A set may be finite or infinite. N is the set of counting numbers and there are infinitely many of these. Since it is impossible to list all of them

1

we use the "..." notation, which means "and so on." In other words, the three dots mean that we are to continue counting according to the pattern already begun. If we wrote

$$E = \{2, 4, 6, \ldots\}$$

we could mean "the set of all positive even integers."

A set may even have no members at all. This set is called the *empty set* or the *null set,* and is denoted by the symbol "$\varnothing$."

Another way of describing a set is by the "rule" method. In this method we use a rule to decide whether an object belongs to a set. For example, S might be the set of all persons in this room who have blond hair. Now to decide whether or not a given object x is in S, apply the rule. Is x a person in this room who has blond hair? If so, $x \in S$. If not, $x \notin S$. We can write this more briefly as follows:

$$S = \{x \,|\, x \text{ is a person in this room having blond hair}\}$$

The symbol "$|$" is read "such that." Note that x is merely a symbol standing for a member of the set, just as you used x in algebra to stand for a number. This should not be interpreted as saying that the *letter x* belongs to the set.

$$N = \{x \,|\, x \text{ is a counting number}\}$$
$$C = \{x \,|\, x \text{ is the capital of Texas}\}$$
$$\varnothing = \{x \,|\, x \text{ is a living dinosaur}\}$$

Consider the set $D = \{1, 2, 3\}$. This is not the same set as N, which is the set of *all* the counting numbers, yet every element of D is a counting number. We say that D is a *subset* of N.

A set A is a subset of a set B if every element of A is an element of B also. We write this symbolically $A \subset B$. The symbol "$\subset$" is read "is a subset of." Note that since every element of the set A is certainly found in A, then $A \subset A$.*

As before, a slash through the symbol denotes negation, and the symbol $\not\subset$ means "is not a subset of."

Thus, $D \subset N$, but $N \not\subset D$. N is not a subset of D because there are elements in N (5, for instance) which are not in D.

Two sets are equal if they contain exactly the same elements. Thus, $A = B$ means that A and B are merely two names for the same set. Every

---

*Some authors reserve the symbol "$\subset$" for the case in which the subset is not the same as the "parent" set, and use the symbol "$\subseteq$" for the case in which the two might be the same set. We will not make this distinction.

element of A is an element of B ($A \subset B$), and every element of B is an element of A ($B \subset A$).

Thus, if

$$A = \{1, 2, 3, 4\}$$
$$B = \{4, 3, 2, 1\}$$

then $A = B$.

Note that since a set is simply a collection of objects, no special order is implied. The elements of N were listed in order for convenience. (If we write $N = \{3, 2, 1, 4, \ldots\}$, it is not clear what pattern "$\ldots$" indicates.)

2. Operations on Sets

Union

If we take two numbers, a and b, we can add them together to get a third number, their sum $a + b$. This is called a binary operation. Subtraction, multiplication, and division are other binary operations on numbers.

We can also combine two sets in various ways to get a third set. One such operation is called *set union*. Just as we use the symbol "$+$" for the operation of addition on numbers, we use the symbol "$\cup$" for the operation of union on sets. The union of two sets A and B is the set of all elements which are found in A or in B or in both. This new set is denoted by the symbol $A \cup B$. In set notation,

$$A \cup B = \{x \,|\, x \in A \text{ or } x \in B \text{ or both}\}$$

For example, if

$$R = \{1, 2, 3, 4\}$$
$$S = \{3, 4, 5, 6\}$$

then $\qquad\qquad R \cup S = \{1, 2, 3, 4, 5, 6\}$

If $\qquad\quad X = \{x \,|\, x \text{ is a letter in the word "apple"}\}$
$$Y = \{x \,|\, x \text{ is a letter in the word "cat"}\}$$

then $\qquad\qquad X \cup Y = \{a, p, l, e, c, t\}$

Note that we did not list 3 and 4 twice in $R \cup S$, nor did we list a and p twice in $X \cup Y$. This would have been redundant. If the letter p is in the set, we say this by listing it, and there is no need to say it twice.

Intersection

Another operation on sets is called *set intersection*. We use the symbol "∩" for intersection and the intersection of two sets A and B is denoted by the symbol $A \cap B$. It is a set containing all those elements that are in *both* A and B.

$$A \cap B = \{x \mid x \in A \text{ and } x \in B\}$$

For example, in the sets

$$R = \{1, 2, 3, 4\}$$
$$S = \{3, 4, 5, 6\}$$

$R \cap S = \{3, 4\}$ since only the elements 3 and 4 are found in both R and S. In the same way, for the sets

$$X = \{x \mid x \text{ is a letter in the word "apple"}\}$$
$$Y = \{x \mid x \text{ is a letter in the word "cat"}\}$$

$X \cap Y = \{a\}$

If two sets have no elements in common, then they are said to be *disjoint* and their intersection is the empty set.

If $T = \{x \mid x \text{ is a triangle}\}$

$Q = \{x \mid x \text{ is a quadrilateral}\}$

then T and Q are disjoint sets and

$$T \cap Q = \varnothing$$

Venn Diagrams

We often draw diagrams to illustrate operations on sets and the relations between sets. In Figure 1-1, the points inside and on the boundary of the circle labeled A represent set A, and the points inside and on the boundary of the circle labeled B represent the set B.

Since we have drawn A and B as overlapping sets, we infer that in this case A and B have some elements in common. They are not disjoint sets. If they were disjoint, we would draw the diagram in Figure 1-2. $A \cup B$ is still the shaded area, but $A \cap B$ is the empty set.

If A is a subset of B, then every element of A is contained in B and we picture the situation in Figure 1-3.

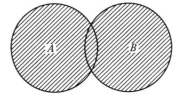

$A \cup B$ is represented by the shaded area.

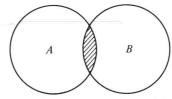

This shaded area represents the set $A \cap B$.

Figure 1-1

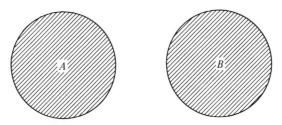

A and B are disjoint sets.

Figure 1-2

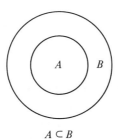

$A \subset B$

Figure 1-3

In this case, it is not hard to see that $A \cup B = B$ and $A \cap B = A$.

Such diagrams are called *Venn diagrams*. Although we have used circles here, we could just as easily have used rectangles, squares, or other shapes to represent our sets.

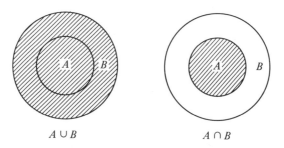

$A \cup B$ $A \cap B$

Figure 1-4

Cartesian Product

The idea of an *ordered pair* is an important and useful one in mathematics. As the name implies, an ordered pair consists of two elements taken in some specified order. For example, your name—leaving out the middle name or initial—is an ordered pair of words. You have a first and a last name. Ordered pairs are particularly useful in describing locations. Your address is probably an ordered pair consisting of a number and a street name.

Symbolically, we write an ordered pair (a, b), where a stands for the first element in the ordered pair, and b represents the second. This is *not* the same thing as the *set* $\{a, b\}$, since there is no order implied in the listing of elements of a set. The set $\{a, b\}$ is exactly the same as the set $\{b, a\}$ since they contain the same elements. The ordered pair (a, b), however, is usually *not* the same as the ordered pair (b, a), since in one case a is the first element and in the second case b is first. To say that two ordered pairs are *equal* means that they have the same first element and the same second element. Thus,

$$(a, b) = (x, y)$$

means $a = x$ and $b = y$.

Since an ordered pair is composed of two elements, there are two sets involved in describing a collection of ordered pairs.

Suppose A and B are two non-empty sets. Then the *Cartesian product* of the two sets A and B, written $A \times B$, (read "A cross B") is defined to be the set of all ordered pairs whose first element comes from A and whose second element comes from B. In symbols

$$A \times B = \{(a, b) \,|\, a \in A \text{ and } b \in B\}$$

If either A or B is the empty set, then $A \times B = \varnothing$ also.

The name "Cartesian product" honors the French mathematician René Descartes who described the position of points in a plane by ordered pairs of real numbers.

If $A = \{1, 2, 3\}$ and $B = \{a, b\}$, then

$$A \times B = \{(1, a), (1, b), (2, a), (2, b), (3, a), (3, b)\}$$

Clearly $B \times A$ is a completely different set.

$$B \times A = \{(a, 1), (b, 1), (a, 2), (b, 2), (a, 3), (b, 3)\}$$

The sets A and B do not have to be different. For example, $A \times A$ is the Cartesian product of A with itself, and

$$A \times A = \{(1, 1), (1, 2), (1, 3), (2, 1), (2, 2), (2, 3), (3, 1), (3, 2), (3, 3)\}$$

To picture the elements of the Cartesian product in a more graphic form, we can use the same scheme used by map makers. Draw some horizontal and vertical lines. Label the vertical lines with elements of A, and the horizontal lines with elements of B. Each intersection can then be labeled with the appropriate ordered pair.

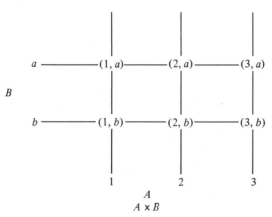

Figure 1-5

One-to-One Correspondence

Two sets A and B are said to be in *one-to-one correspondence* if we can find a way of matching their elements so that every element of A is paired with exactly one element of B and every element of B is paired with exactly one element of A.

For example, if

$$A = \{1, 2, 3, 4\}$$
$$B = \{a, b, c, d\}$$

then A and B are in one-to-one correspondence since we can pair elements like this

$$1 \longleftrightarrow a \quad 2 \longleftrightarrow b \quad 3 \longleftrightarrow c \quad 4 \longleftrightarrow d$$

Note that in this pairing (called a *correspondence*) every element of A is used exactly once and every element of B is used exactly once. This is, of course, not the only way that elements of A can be matched with elements of B. Another correspondence would be

$$1 \longleftrightarrow d \quad 2 \longleftrightarrow c \quad 3 \longleftrightarrow b \quad 4 \longleftrightarrow a$$

If two finite sets are in one-to-one correspondence then they have the same number of elements.

Exercise 1.2

1. Let Q be the set of all quadrilaterals. Indicate by symbolic notation (i.e., $a \in A$, $b \notin A$) whether each of the following elements is or is not a member of Q.

    (a) a square, s (b) a triangle, t

    (c) a parallelogram, p (d) a rectangle, r

    (e) a pentagon, a (f) a circle, c

2. Let Q be the set of quadrilaterals, S the set of squares, R the set of rectangles, and P the set of parallelograms. Which of these sets are related by the relation "is a subset of"? Use symbolic notation ($A \subset B$) to denote this for every pair of sets for which it is true.

3. Find the union and the intersection of each of the following pairs of sets:

    (a) $E = \{2, 4, 6, \ldots\}$; $O = \{1, 3, 5, \ldots\}$

    (b) $B = \{x \mid x$ is a boy in the freshman class$\}$;
    $G = \{x \mid x$ is a girl in the freshman class$\}$

    (c) $A = \{1, 2, 3, 4, 5\}$; $B = \{1, 3, 5\}$

    (d) $R = \{x \mid x$ is a rectangle$\}$;
    $S = \{x \mid x$ is a square$\}$

    (e) $E = \{2, 4, 6, 8, \ldots\}$; $F = \{3, 6, 9, 12, \ldots\}$

4. Draw a Venn diagram illustrating the relationship between each pair of sets in problem 3.

5. Draw a Venn diagram illustrating the relationship between the set Q of quadrilaterals and the set R of rectangles; between the set T of

triangles and the set S of squares; between the set R of rectangles and the set H of rhombuses.

6. Let $A = \{0, 1\}$; $B = \{$red, white, blue$\}$. List the elements of $A \times B$; of $B \times A$. Illustrate these sets with a diagram of vertical and horizontal lines as in the text.

7. If the set A has 4 elements and the set B has 3 elements, how many elements are in $A \times B$? In $B \times A$? In general, if A has n elements and B has m elements, how many elements are in $A \times B$? In $B \times A$?

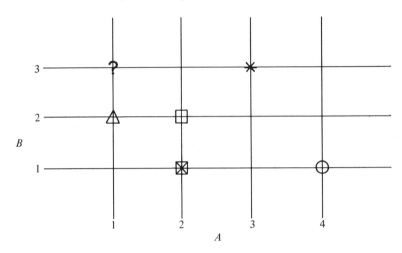

8. The intersections of the lines above represent elements of $A \times B$. Describe by giving an ordered pair of numbers the location of $\triangle$, of ?, of $\square$, of $\boxtimes$, of $*$, of $\bigcirc$.

9. If $A \times B = B \times A$, what can you infer about the sets A and B?

3. Relations

Picture a small child playing with a set of colored blocks. He is sorting them according to color, the red blocks in one pile, the blue ones in another and so on. When he has finished, the set of blocks has been sorted into disjoint subsets, each subset characterized by its color. Two blocks are in the same subset if and only if they have the same color. If two blocks have the same color then we say they are *related* by this property. If they have different colors they are not related. This relation of having the same color is an example of an *equivalence relation*. We call it that because if two blocks have the same color we consider them to be equivalent as far as color is concerned. If we say to the child, "Hand me a red block," it doesn't matter which red block he hands us.

There are other equivalence relations which we could consider on the set of blocks. The child might have sorted them according to shape, rectangular blocks in one pile, triangular-shaped ones in another and so on.

A characteristic of an equivalence relation is that it subdivides the set on which it is defined into disjoint subsets. Each subset is called an *equivalence class*, since any two objects from this set are equivalent.

What characterizes an equivalence relation? Well, first we note that every block in the set is related to itself, since it is the same color as itself. We call this the *reflexive property*. Moreover, if we have two blocks, say *a* and *b*, with *a* the same color as *b*, it is certainly true that *b* is the same color as *a*. This illustrates the *symmetric property*. Finally, if we have three blocks, *a*, *b*, and *c*, and if *a* is the same color as *b* and *b* is the same color as *c*, then we can conclude that *a* is the same color as *c*. This is called the *transitive property*.

We can give a more formal definition of these three properties. Let us call our set *S*, and let the letters *a*, *b*, *c* stand for *any* elements of *S*. (This does not imply that *S* has only three members, nor does it mean that *S* must have at least three members.) Let $\circledR$ be the symbol for "relation." We will write $a \circledR b$, meaning "*a* is related to *b*." In the above example $\circledR$ was "is the same color as."

$\circledR$ is an equivalence relation provided it has all three of the following properties:

(i)   The reflexive property. For every $a \in S$, $a \circledR a$.
(ii) The symmetric property. If $a \circledR b$, then $b \circledR a$.
(iii) The transitive property. If $a \circledR b$ and $b \circledR c$, then $a \circledR c$.

We will call these the *R*, *S*, *T* properties.

Property (iii) does not demand that the set *S* have at least three members, because the elements *a*, *b*, and *c* need not be different. For example, *a* and *b* could be two names for the same object. The relation "is the same color as" is still an equivalence relation, even if the child has only two blocks (or even one!) in his set.

Not every relation is an equivalence relation. Suppose the child is arranging the blocks according to size. He lines them up in a row with the smallest one first, followed by the next larger and so on. The first block *is smaller than* the second; the second *is smaller than* the third. The relation $\circledR$ here is "is smaller than."

Clearly this relation has the transitive property. If *a* is smaller than *b* and *b* is smaller than *c*, then *a* is smaller than *c*. However, this relation is neither reflexive nor symmetric. No block is smaller than itself, and if *a* is smaller than *b*, clearly it is impossible for *b* to be smaller than *a*. This is an example of a very useful relation called an *ordering* relation.

We have looked at a number of examples of relations but have not yet said just what a relation *is*. In each case note that certain *pairs* of objects were said to be "related." Moreover, this was an *ordered* pair, since *a* might be related to *b*, but *b* fail to be related to *a*. The mathematician defines a relation very loosely as "a set of ordered pairs of objects from some set *S*"—that is, a subset of $S \times S$. Thus Ⓡ is a *set* of ordered pairs, and if the ordered pair (a, b) is in the set Ⓡ, then we say that a Ⓡ b (a is related to b.)

As an example, consider the set of ordered pairs

$$Ⓡ = \{(2, 1), (3, 1), (4, 1), (3, 2), (4, 2), (4, 3)\}$$

If we take the set *S* to be $\{1, 2, 3, 4\}$ then this is a subset of $S \times S$, hence is a relation. Note that in every ordered pair the first element is larger than the second. Another way of describing Ⓡ would be as the relation "is greater than" on the set $\{1, 2, 3, 4\}$.

In what follows we will be looking at many relations on sets of geometric figures, such as the relation "is parallel to" on the set of lines, and the relation "is congruent to" on the set of triangles. Some of these will be equivalence relations and some will not. You will frequently be asked which of the *R, S, T* properties a relation has.

Exercise 1.3

Explain why each of the following is an equivalence relation and describe a typical equivalence class.

1. The relation "is the same age as" on the set of students in this classroom.

2. The relation "has the same surname as" on the set of people in this city.

3. The relation "has the same author as" on the set of books in the library.

4. The relation "has the same number of sides as" on the set of polygons.

5. The relation "is in a one-to-one correspondence with" on the set of finite sets.

Which of the *R, S, T* properties do the following relations have?

6. The relation "is taller than" on the set of people in this classroom.

7. The relation "divides" on the set of counting numbers. (We say that *a* divides *b* if $b \div a$ is an integer. For example, 2 divides 4, but 2 does not divide 3.)

8. The relation "is the mother of" on the set of people.

9. The relation "is the brother of" on the set of people.

10. The relation "is a subset of" on the collection of all subsets of

$$S = \{1, 2, 3, 4, 5\}.$$

11. The relation "is supplementary to" on the set of angles. (Two angles are supplementary if the sum of their measures is 180°.)

12. The relation "is an antonym of" on the set of words in the English language. (Two words are antonyms if they are opposite in meaning. For example, *good* and *bad* are antonyms.)

13. Let ⓡ = {(1, 1), (2, 2), (3, 3), (4, 4)}. This is a subset of $S \times S$ where $S = \{1, 2, 3, 4\}$. What is another way of describing the relation ⓡ?

4. Logic

Introduction

"That proves it." "That is no proof." We hear phrases like these every day. What is a proof? More specifically, what is a mathematical proof? Every textbook in mathematics contains proofs; in fact, many have little else. If we are to read a mathematics textbook with understanding, we must be familiar with the language of proof, and we need some knowledge of the logical principles involved.

True-False Statements

A *true-false statement* (abbreviated t-f statement) is a declarative sentence that is either true or false, but not both. We say that every such statement has a truth value which may be "true" or "false."

The following are t-f statements:

1. Seven is an odd number.

2. Pigs can fly.

These are not t-f statements.

3. Who are you?

4. This statement is false.

If sentence 4 is true, it is false and if it is false, it is true; therefore, we would not consider it to be a t-f statement.

We will let small letters, $p, q, r, s, \ldots$ stand for t-f statements. If p is any t-f statement, then the statement "p is false" is also a t-f statement called the *negation* or the *denial* of p, and is denoted by the symbol $\sim p$.

Set Theory and Logic

(Read "not p.") Clearly, if p is true, then $\sim p$ is false, and if p is false, $\sim p$ is true; p and $\sim p$ have opposite truth values.

For example, if p is the statement "The sun is shining" then $\sim p$ is the statement "The sun is not shining." The student should be cautioned, however, that the negation of a statement is not always so simple to state. The two statements

Not all cats have blue eyes.

All cats do not have blue eyes.

for example, do not say the same thing at all! Only one of them is the negation of "All cats have blue eyes." (The first one.) If it is not true that all cats have blue eyes, then *some* cats (at least one) do not have blue eyes. This is what the first statement says. The second, however, asserts that *no* cats have blue eyes. However, $\sim p$ can always be expressed in the form "It is not the case that . . . ," or "It is false that . . ."

Compound t-f Statements

We can combine two t-f statements to make a new one by using the connectives "and" and "or." If p is the statement "It is raining" and q the statement "The sun is shining," we obtain a new statement by writing "It is raining *and* the sun is shining." If we use the symbol " $\wedge$ " for the word "and," our new statement may be written symbolically as "$p \wedge q$." This is called the *conjunction* of p and q.

The truth value of $p \wedge q$ clearly depends on the truth values of p and q. If *both* p and q are true, then $p \wedge q$ is true. If either p or q (or both) is

p	q	$p \wedge q$
T	T	T
T	F	F
F	T	F
F	F	F

Table 1-1

false, then $p \wedge q$ is false. We can describe this relationship in a *truth table*.

Note the similarity between the symbol "$\wedge$" for conjunction and the symbol "$\cap$" for set intersection. Since $A \cap B$ is the set of all elements which are in A *and* in B, we could write

$$A \cap B = \{x \,|\, x \in A \wedge x \in B\}$$

We can also connect the two statements with the word "or." "It is raining *or* the sun is shining." The symbol "$\vee$" stands for the word "or" and our new statement may be written "$p \vee q$." It is called the *disjunction* of p and q.

Here we run into the fact that "or" may have two meanings in English. It can mean "one or the other or both" (the inclusive or), or it can mean "one or the other, but not both" (the exclusive or). For example, consider the statement "I will make an A or a B in this course." Obviously, you cannot do both, so this is an example of the exclusive or. In mathematics, however, "or" usually means the inclusive or and this will be our interpretation of "$\vee$." Thus $p \vee q$ will be true provided at least one of the statements p, q is true. Its truth table looks like this:

p	q	$p \vee q$
T	T	T
T	F	T
F	T	T
F	F	F

Table 1-2

Again note the similarity between the symbol "$\vee$" for disjunction and the symbol "$\cup$" for set union. Since $A \cup B$ is the set of elements which are in A or in B or in both,

$$A \cup B = \{x \,|\, x \in A \vee x \in B\}$$

The Language of Proof

A *theorem* is a t-f statement of the form "if p, then q," where p and q are t-f statements. A statement of this form is called a *conditional* or an *implication*. We often say that p "implies" q and write this $p \Rightarrow q$.

In a theorem, the first statement, p, is called the *hypothesis* and the second statement, q, is called the *conclusion*. You may object that many theorems do not read like this at all, and that is true. However, if you examine any theorem closely you will see that it can be restated in the "if . . . then" form without changing its meaning. For example, consider the statement "All equilateral triangles are isosceles." An equivalent way of stating this would be, "If a triangle is equilateral, then it is isosceles."

The conditional $p \Rightarrow q$ may be worded in a number of different but equivalent ways. The phrase *only if* in a theorem frequently causes confusion, probably because in everyday English "if" and "only if" are often used interchangeably. In mathematics, however, their meanings are quite different. "If p then q" is equivalent to saying "p only if q."

For example,

p: A triangle is equilateral.

q: A triangle is isosceles.

The conditional $p \Rightarrow q$ might be worded
   (1) If a triangle is equilateral, then it is isosceles.
or, equivalently
   (2) A triangle is equilateral only if it is isosceles.

Two other words frequently encountered in the statement of theorems are *necessary* and *sufficient*.

The following are all equivalent ways of expressing $p \Rightarrow q$:
   (1) if p then q
   (2) p only if q
   (3) p is sufficient for q
   (4) q is necessary for p

Thus, we might say
   (3) The fact that a triangle is equilateral is a sufficient condition for it to be isosceles.
   (4) The fact that a triangle is isosceles is a necessary condition for it to be equilateral.

If a theorem is of the form $p \Rightarrow q$, the statement $q \Rightarrow p$ that we get by interchanging hypothesis and conclusion is called its *converse*. The converse of the statement above would be:

If a triangle is isosceles, then it is equilateral.

This statement is not true since we can find isosceles triangles which are not equilateral. (A triangle is isosceles if at least two sides are congruent. It is equilateral if all three sides are congruent.)

Thus a theorem may be true and its converse false. If a theorem and its converse are *both* true, i.e., $p \Rightarrow q$ and $q \Rightarrow p$ are both true statements, then we say that p and q are equivalent statements, and the theorem and its converse are often written together.

"p if and only if q" (abbreviated p iff q)

or   "p is necessary and sufficient for q"

In symbols we write $p \Leftrightarrow q$.

For example, since

　　(1)　If a triangle is equilateral, then it is equiangular

and

　　(2)　If a triangle is equiangular, then it is equilateral

are both true statements, we can combine the two into one.

A triangle is equilateral if and only if it is equiangular

or

A necessary and sufficient condition for a triangle to be equilateral is that it be equiangular.

An "if and only if" theorem is really two theorems in one, and proving such a theorem involves two steps. First we must prove that $p \Rightarrow q$ is true, then that the converse $q \Rightarrow p$ is true also.

A *definition* is always understood to be an "if and only if" statement, although it is seldom phrased this way. Consider, for example, the definition of a right angle, "A right angle is an angle whose measure is 90°." This means (1) If this is a right angle, then its measure is 90°, and (2) If this angle has measure 90°, then it is a right angle.

Given a theorem of the form $p \Rightarrow q$, the statement $\sim q \Rightarrow \sim p$ (the denial of q implies the denial of p) is called the *contrapositive* of the theorem. The contrapositive is a very interesting and useful statement because it is logically equivalent to the statement $p \Rightarrow q$. By this we mean they have the same truth values. If $p \Rightarrow q$ is true, then $\sim q \Rightarrow \sim p$ is true, and if $p \Rightarrow q$ is false, then $\sim q \Rightarrow \sim p$ is false.

If you have proved that a theorem is true, then you know that its contrapositive is true also, and it can be stated as a theorem as well.

For example,

　　$p \Rightarrow q$:　If a triangle is equilateral, then it is isosceles.

　　$\sim q \Rightarrow \sim p$:　If a triangle is not isosceles, then it is not equilateral.

The statement $\sim p \Rightarrow \sim q$ is called the *inverse* of $p \Rightarrow q$ and it is logically equivalent to the converse, $q \Rightarrow p$.

Example:

$p \Rightarrow q$:   If two angles are right angles, then their measures are equal. (True)

$q \Rightarrow p$ (converse):   If the measures of two angles are equal, then they are right angles. (False)

$\sim q \Rightarrow \sim p$ (contrapositive):   If the measures of two angles are not equal, then they are not right angles. (True)

$\sim p \Rightarrow \sim q$ (inverse):   If two angles are not right angles, then their measures are not equal. (False)

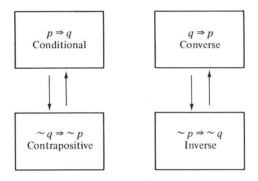

Figure 1-6

An important part of most theorems is a quantifying phrase such as "for all," "for every," "for each." These phrases are used interchangeably. Most theorems can be rewritten in the form, "for all objects x, if so-and-so is true about x, then something else is true about x."

For example, consider the theorem, "All equilateral triangles are isosceles." This means, for all (or for each, or for every) triangles x, if x is equilateral, then x is isosceles.

The phrase "for all" is called a *universal quantifier*. The theorem asserts that every object in the set of equilateral triangles is in the set of isosceles triangles.

In many theorems the universal quantifier is not explicitly stated, but is implied. For example, we might say, "Vertical angles are congruent." We understand this to apply to *all* pairs of vertical angles, and the theorem could be stated "For all angles x and y, if x and y are vertical angles, then x is congruent to y."

A different quantifier is the *existential quantifier*, "there exists," "there is" or "some." These have the same meaning. We understand "some" to mean *at least one* and "there exists" asserts the existence of *at least one.* For example, we might say, "Some isosceles triangles are equilateral" or "There exists an isosceles triangle which is equilateral." This means there is at least one element in the set of isosceles triangles which is also in the set of equilateral triangles.

This throws some light on the question "When is an example a proof?" It is not hard to see that if a theorem says something about *all* objects in some infinite set (say the set of all equilateral triangles) then showing the theorem holds for one specific object in that set will prove nothing. In this case an example can never be a proof.

On the other hand, suppose the theorem uses the existential quantifier "there exists" or "some." Since we interpret this to mean we must show that the theorem holds for at least one object in the set, in this case an example will be a proof. (To prove that "Some isosceles triangles are equilateral" we can display an example of an isosceles triangle which is also equilateral.)

Methods of Proof

We will frequently need to *prove* theorems, that is, to show that they are true statements. The truth value of $p \Rightarrow q$ will, of course, depend on the truth values of the statements p and q. The statement $p \Rightarrow q$ is said to be false if and only if p is true but q is false; that is, the hypothesis is true but the conclusion is false. For all other combinations of truth values of p and q, $p \Rightarrow q$ is true. The truth table for $p \Rightarrow q$ looks like this:

p	q	$p \Rightarrow q$
T	T	T
T	F	F
F	T	T
F	F	T

Table 1-3

According to the rules of logic, then, if the hypothesis is false, the theorem is true no matter what the conclusion says. Sometimes if the hypothesis is false we say the theorem is trivially true. In actual practice, of course, these cases seldom arise, so we will usually assume that the hypothesis of our theorem is a true statement.

If the hypothesis is true, then the theorem is true if and only if the conclusion is true. Thus in the proof of a theorem we usually try to show that the conclusion is true.

A proof may be direct or indirect. In a direct proof we proceed step by step, using the hypothesis and any axioms and previously proved theorems which we need until we arrive at the conclusion as a consequence of our chain of reasoning. Of course, a plan of attack is essential. Writing down statements at random, even if they are true, usually will not lead to the desired conclusion.

Planning a direct proof can be compared to planning a trip by car. If you want to take a trip to some place you have never visited before, you don't jump in the car and drive around at random hoping you will run across your destination. If you are wise, you get a road map, find your present location on the map, look for the place you are going, then plan a route.

In planning a proof you look to see where you are, i.e., what facts, axioms, and theorems you can use in your proof; find where you want to go, that is, what you are trying to prove; and plan a method of attack.

Indirect proof, also called proof by contradiction or *reductio ad absurdum*, is a very common method of proof. In an indirect proof, we assume the theorem to be false; that is, we assume the hypothesis is true and the conclusion is false. Then we explore the consequences of these assumptions. Suppose this leads us to the conclusion that some statement, say *r*, which we know to be true, is false. The statement *r* might be a part of the hypothesis, an axiom or a previously proved theorem. Since *r* cannot be both true and false, our assumption that the theorem is false has led to a contradiction. Therefore, our theorem must be true. We will see many examples of indirect proof in the chapters that follow.

Conjecture

When a mathematician notices that in a number of examples a certain relationship holds, he suspects that it holds for all cases. He might state such an educated guess as a tentative theorem, or *conjecture*. He then tries to prove his conjecture. If he succeeds, of course, it is a conjecture no longer, but a full-fledged theorem. If he is not able to prove his conjecture, he may suspect that there is some example for which it does not hold. Such an example is called a *counterexample*. The existence of only *one* counterexample is enough to prove the conjecture false, since to be true the conjecture had to be true for *all* cases.

A famous case is *Goldbach's Conjecture*. This conjecture states that every even integer greater than 2 can be written as the sum of two positive primes. (One is not considered to be prime.) For example,

$$4 = 2 + 2 \quad\quad 6 = 3 + 3 \quad\quad 8 = 5 + 3 \quad\quad 10 = 7 + 3$$

This conjecture has been observed to be true for many examples, but it has never been proved to be true for *all* cases. Moreover, no counter-example has ever been found.

Exercise 1.4

1. In a certain town the barber shaves every man who does not shave himself. Is this sentence a t-f statement?

2. A prisoner was told he could choose the method of his execution. He was to make a statement. If he made a true statement he was to be shot; if false he was to be hanged. The prisoner replied, "Well, I'll be hanged." Was this a t-f statement?

3. Restate the following in "if . . . then" form. Underline the hypothesis once, the conclusion twice.

   (a)   All right angles are congruent.

   (b)   Congruent triangles are similar.

   (c)   Base angles of an isosceles triangle are congruent.

   (d)   The sum of the measures of the angles of a triangle is $180°$.

   (e)   The sum of two even integers is an even integer.

4. State the converse of each of the theorems in problem 3.

5. State the contrapositive of each of the theorems in problem 3.

6. What is the converse of the converse of a theorem?

7. Let p: Two angles are vertical angles.

   q: Two angles are congruent.

   State the conditional $p \Rightarrow q$ in the form (a) if p then q, (b) p only if q, (c) p is sufficient for q, and (d) q is necessary for p.

8. Let p: A number is prime and greater than two.

   q: A number is odd.

   State the conditional $p \Rightarrow q$ in the form (a) if p then q, (b) p only if q, (c) p is sufficient for q, and (d) q is necessary for p.

9. State the converse, the contrapositive, and the inverse of "If wishes were horses, then beggars would ride."

10. State the converse, the contrapositive, and the inverse of "If an integer is divisible by 6 then it is divisible by 2." Which of these are true statements?

11. What is the contrapositive of the converse of $p \Rightarrow q$?

12. The following statements are equivalent. Write the statement $p \Leftrightarrow q$ using the phrase "if and only if"; using the phrase "necessary and sufficient."

    (a) p: A triangle is isosceles.
       q: A triangle has two equal angles.

    (b) p: A number is divisible by 6.
       q: A number is divisible by both 2 and 3.

    (c) p: A quadrilateral is a parallelogram.
       q: A quadrilateral has both pairs of opposite sides congruent.

    (d) p: A number is divisible by 5.
       q: The units digit of a number is either 0 or 5.

13. Prove that the following conjectures are false by finding a counterexample.

    (a) For all integers x, if x is divisible by 3, then x is divisible by 9.

    (b) For all triangles x, if x is isosceles, then x is equilateral.

    (c) For all integers x, if x is greater than 2, then x can be written as the sum of two primes.

    (d) For all integers x, if x is prime then x is of the form

    $$6n + 1 \quad \text{or} \quad 6n - 1$$

New Terms Found in This Chapter

Term	Section	Term	Section
set	1	negation (denial)	4
element of a set	1	conjunction	4
empty set	1	truth table	4
subset	1	disjunction	4
set union	2	conditional (implication)	4
set intersection	2	hypothesis	4
Venn diagram	2	conclusion	4
ordered pair	2	converse	4
Cartesian product	2	contrapositive	4
equivalence relation	3	inverse	4
equivalence class	3	universal quantifier	4
reflexive property	3	existential quantifier	4
symmetric property	3	proof by contradiction	
transitive property	3	(*reductio ad absurdum*)	4
relation	3	conjecture	4
true-false statement	4	counterexample	4

2: Sets of Points in Space

1. Definitions

One might think that in any mathematical discussion the first thing to do would be to define all terms very carefully. A little thought will show that this is impossible. New words can be defined only in terms of words already defined. Thus, in attempting to define every term either we arrive at some "first" terms which we cannot define or else we go in circles, defining A in terms of B, B in terms of C, and C in terms of A, and end in a circumlocution that defines nothing.

Therefore, in geometry we will deliberately leave some terms undefined and use these "primitive" terms as a basis for defining others. Some terms that usually remain undefined are "point," "line," and "plane" and in this book we will follow the standard pattern.

This point of view is a fairly modern one. Euclid, for example, attempted to define all of his terms. His idea seemed to be to define geometric notions such as *point* and *line* using non-geometric terms. In Book I of Euclid's *Elements*, Definition 1 states, "A point is that which has no part." We might ask whether this definition makes clear what a point is. If you did not already have an intuitive notion of a point, this definition would not be very helpful. You might infer from this definition that a point was nothing at all, since it has no parts.

Definition 2 says that, "A line is breadthless length." One wonders why Euclid did not define "length" and "breadth" since these are also geometric ideas. He seems to be defining a line in terms of two other terms, length and breadth, which are perhaps not so intuitively simple as the idea being defined.

When we do define new terms, we will be very careful and precise. In mathematics a definition is not a synonym but a yard stick, a criterion that we use to judge whether an object can be described by the term we are defining.

For example, we say that two lines are *parallel* if they lie in the same plane and their intersection is the empty set. Note that the term "parallel" is defined using the undefined terms "line" and "plane" and the notion of set intersection.

Given two lines, we apply this definition. Are they in the same plane? Is their intersection empty? If these questions can be answered "yes," they must be parallel. If either of these qualifications is not met, then we conclude that the lines are not parallel.

2. Points, Lines, and Planes

Although "point," "line," and "plane" will be undefined terms, we can think of a point intuitively as a location in space, perhaps represented by a chalk dot on the blackboard, or a pencil dot on a piece of paper. We will use small letters, p, q, r, and so on, to name points.

We will understand a "line" to mean a straight line, one that stretches out infinitely far in each direction and we will indicate this by a line drawn on our paper with arrows at either end. We will use capital letters such as L_1, L_2 to indicate lines. In set theory we used small letters for elements of sets and capital letters for sets. Since in geometry all our sets are sets of points, our notation for point and line is consistent. We may also name lines by giving two points on the line.

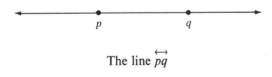

The line $\overleftrightarrow{pq}$

Figure 2-1

Thus, if p and q are points on the line, we can refer to the line by the symbol $\overleftrightarrow{pq}$ or $\overleftrightarrow{qp}$. Intuitively, we can think of a line as a very thin wire stretched so tightly that it does not sag at all, or as the edge of a ruler. Actually, these models represent only small parts of lines, since a line is considered to be infinite in extent.

A "plane" may be thought of intuitively as a flat surface stretching out infinitely far in every direction. Thus, a table top or a sheet of paper may be thought of as a very small part of a plane. We will use capital Greek letters such as Π, Σ to name planes.

The student should realize that the dot on his paper, the pencil mark drawn with a straight-edge and the top of his desk are not a "point," a "line," or a "plane" any more than the numeral "2" is the *number* two. We emphasize again that "point," "line," and "plane" are *undefined*. The intuitive concepts of point, line, and plane we have described are merely one interpretation of these terms and not necessarily the only interpretation. We will go into this in more detail later in the section on mathematical models.

In studying geometry we will use the language of set theory. Thus, "space" will be the set of all points, while geometric figures such as lines, planes, and triangles will be certain subsets of space. At the same time, we will not abandon the traditional language of geometry. We will still say that a line "passes through" a point, or a point "lies on" a line. In terms of set theory this means that the point is an element of the set of points which is the line.

Often the old phrases will be very similar to the new terms of set theory. We commonly say two lines "intersect" when the intersection of the two sets is not empty.

3. Separation Properties

A point p on a line L separates the line into three disjoint sets of points. One of these sets is the point p itself. The other two subsets of L are called *half-lines* and consist of the points on either side of p.

A *ray* is a subset of a line consisting of a point and one of the half-lines determined by it.

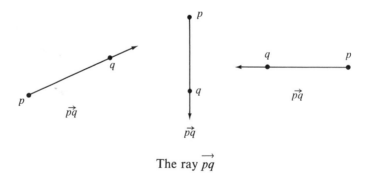

The ray $\overrightarrow{pq}$

Figure 2-2

If p is the separating point, we will call p the *end-point* of the ray. If q is any point on the half-line, then we can describe the ray by the symbol $\overrightarrow{pq}$. This symbol should not be interpreted as saying that the ray points to the right, or that q is to the right of p. A ray can point in any direction, and if p is its end-point and q some other point on the ray, we can call it $\overrightarrow{pq}$. We call two rays *opposite rays* if their union is a line and their intersection is a single point, their common end-point.

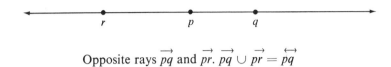

Opposite rays $\overrightarrow{pq}$ and $\overrightarrow{pr}$. $\overrightarrow{pq} \cup \overrightarrow{pr} = \overleftrightarrow{pq}$

Figure 2-3

In Figure 2-3 $\overrightarrow{pq}$ and $\overrightarrow{pr}$ are opposite rays. Their union is the line $\overleftrightarrow{pq}$ (or $\overleftrightarrow{rp}$ or $\overleftrightarrow{qr}$. Recall that a line can be named by listing any two points on it with the double arrow above.)

A *line segment* is a subset of a line. It consists of two different points on the line, called end-points, and all the points between them. The notion of "betweenness" is another term which we shall take as undefined. Intuitively, it is clear that if p, q, and r are three points on a line, then one of them lies "between" the other two. If p and q are the end-points of the line segment, we use the symbol $\overline{pq}$ to describe the line segment. We can think of the line segment $\overline{pq}$ as the intersection of two rays, $\overrightarrow{pq}$ and $\overrightarrow{qp}$.

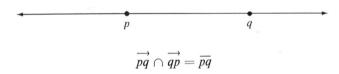

$$\overrightarrow{pq} \cap \overrightarrow{qp} = \overline{pq}$$

Figure 2-4

Just as a point separates a line into three disjoint subsets, so a line separates a plane into three disjoint subsets: the line itself, and two sets of points on either side of the line, called *half-planes*.

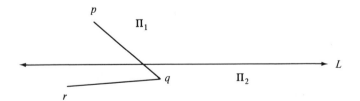

The line L separates the plane into two half planes, Π_1 and Π_2.

Figure 2-5

In Figure 2-5 Π_1 and Π_2 indicate the half-planes determined by the line L. Suppose p is some point in Π_1 and q is some point in Π_2. Note that the intersection of the line segment $\overline{pq}$ with L is non-empty. In fact, the intersection will contain exactly one point. On the other hand, if q and r are both in Π_2, then the line segment $\overline{qr}$ will contain no points of L.

Just as a point separates a line and a line separates a plane, so a plane separates space. These notions of separation are closely allied to the notion of *dimension*. If we think of space as being three dimensional, a plane is two dimensional; a line has one dimension; and a point has zero dimension.

Thus, a one dimensional set, a line, can be separated by a zero dimensional set, a point; a two dimensional set, a plane, can be separated by a one dimensional set; while a three dimensional set, space, can be separated by a two dimensional set.

Exercise 2.3

1. Translate the following sentences into the symbols of set theory:
    (a) Lines L_1 and L_2 do not intersect.
    (b) The line L lies in the plane Π.
    (c) The plane Π contains the line L.
    (d) The point p lies on the line L.
    (e) The line L passes through the point p.
    (f) The point p lies in the plane Π.
    (g) The line L intersects the plane Π in the point p.
    (h) The lines L_1 and L_2 intersect in a single point p.

2. Let the point p be on a line L. What is the union of the two half-lines determined by p? What is the union of the two rays determined by p? Let the line L lie in a plane Π. What is the union of the two half-planes determined by L?

3. Draw a line L and on it label three points p, q, and r, with q between p and r. Describe the following sets of points:

    (a) $\overline{pq} \cap \overline{qr}$ (b) $\overline{pr} \cap \overline{qr}$

    (c) $\overline{pq} \cup \overline{qr}$ (d) $\overline{pq} \cup \overline{pr}$

4.

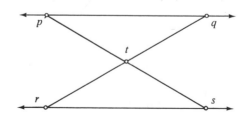

Using the figure above, describe the intersection of the following point sets:

(a) $\overleftrightarrow{pq}$ and $\overleftrightarrow{ts}$ (b) $\overleftrightarrow{pq}$ and $\overline{ts}$

(c) $\overleftrightarrow{pq}$ and $\overleftrightarrow{rs}$ (d) $\overleftrightarrow{pq}$ and $\overrightarrow{pq}$

(e) $\overleftrightarrow{pq}$ and the half-plane above $\overleftrightarrow{rs}$

(f) $\overleftrightarrow{pq}$ and the half-plane above $\overleftrightarrow{rq}$

5. Draw a figure to illustrate each of the following statements:

(a) $\overleftrightarrow{pq} \cap \overleftrightarrow{rs} = \{t\}$ (b) $\overline{pq} \cap \overline{rs} = \overline{rq}$

(c) $\overline{pq} \cap \overline{rs} = \overline{ps}$ (d) $\overrightarrow{pq} \cap \overrightarrow{rs} = \overline{pr}$

(e) $\overline{pq} \cap \overline{rs} = \{p\}$

6. (a) Draw two line segments $\overline{pq}$ and $\overline{rs}$ such that $\overline{pq} \cap \overline{rs} = \varnothing$, but $\overleftrightarrow{pq} \cap \overleftrightarrow{rs}$ is a single point.

(b) Draw two line segments $\overline{pq}$ and $\overline{rs}$ such that $\overline{pq} \cap \overline{rs} = \varnothing$, but $\overleftrightarrow{pq} \cup \overleftrightarrow{rs} = \overleftrightarrow{pq}$.

(c) Draw two line segments $\overline{pq}$ and $\overline{rs}$ such that $\overline{pq} \cap \overline{rs} = \varnothing$, but $\overrightarrow{pq} \cup \overrightarrow{rs} = \overleftrightarrow{pq}$.

7. Let L be a line and on L label four points, a, b, c, and d in that order. Using these points, name two different rays

(a) whose union is L.

(b) whose union contains a, b, c, and d but is not all of L.

(c) whose union does not contain a.

(d) whose union does not contain d.

(e) whose intersection is a single point.

(f) whose intersection is empty.

(g) whose intersection is not empty, but does not contain a.

(h) whose intersection is not empty, but does not contain $\overline{ab}$.

(i) whose intersection contains $\overline{ab}$ but not $\overline{cd}$.

8. Draw a line L and on it label three points p, q, and r, with q between p and r. Tell whether each statement is true or false. If a statement is false, rewrite it making it true.

(a) $\overleftrightarrow{pq} = \overleftrightarrow{qr}$ (b) $\overrightarrow{qr} \cup \overrightarrow{qp} = \overleftrightarrow{pq}$

(c) $\overrightarrow{qr} \cap \overrightarrow{qp} = \{q\}$ (d) $\overrightarrow{pq} = \overrightarrow{qp}$

(e) $\overrightarrow{pq} \cup \overrightarrow{qr} = \overrightarrow{pq}$ (f) $\overrightarrow{pr} \cap \overrightarrow{qp} = \overline{pq}$

(g) $\overrightarrow{qr} \cup \overrightarrow{qp} = \overrightarrow{rp}$ (h) $\{p\} \cup \overline{pq} = \overrightarrow{qp}$

(i) $\overrightarrow{qr} \subset \overrightarrow{pr}$ (j) $\overrightarrow{qr} \subset \overrightarrow{rp}$

(k) $\overrightarrow{qr} \subset \overline{qr}$

9. Let L be a line and on L label the points a, b, c, d, e, and f in that order. Using this figure, find:

(a) $\overline{ac} \cup \overline{ce}$

(b) $\overline{ac} \cup \overline{bd}$

(c) $\overline{ab} \cap \overline{eb}$

(d) $\overrightarrow{de} \cap \overrightarrow{fe}$

(e) $\overrightarrow{de} \cap \overrightarrow{cb}$

(f) $\overleftrightarrow{ab} \cap \overrightarrow{dc}$

(g) $\overline{bc} \cup (\overline{ab} \cap \overline{cd})$

(h) $(\overrightarrow{fe} \cap \overrightarrow{bc}) \cap \overline{ac}$

(i) $(\overrightarrow{de} \cap \overrightarrow{ea}) \cup (\overrightarrow{da} \cap \overrightarrow{cf})$

4. Axioms and Theorems

Just as we cannot hope to define every term, we cannot expect to be able to prove every statement in geometry. Some statements will be accepted without proof and used to prove other statements. The statements which are accepted without proof are called *axioms*, and those which are proved are called *theorems*. You may also be familiar with the term "postulate." Euclid distinguished between "postulates" and other statements he called "common notions." Euclid's postulates were assumptions which were geometric in nature, such as, "All right angles are equal to one another." Euclid considered the common notions to be self-evident truths. The first states, for example, that "Things which are equal to the same thing are also equal to one another." Another says, "The whole is greater than the part." Euclid considered postulates and common notions to be basically different. The reader is asked to grant him the truth of the postulates, although they are not particularly "self-evident." On the other hand, Euclid clearly felt that the common notions were obviously "true." In your experience, for example, you have probably observed that the whole of something is always greater than a part of it.

Today most geometers do not distinguish between postulates and common notions. Both are statements which are assumed to be true without proof, and are called *axioms*. If, in your experience the whole has always been observed to be greater than a part, this does not "prove" that the axiom is "true." You certainly have not observed all possible cases, and so conceivably there might be some instance in which the whole is *not* greater than a part. (Such a phenomenon does in fact occur in the study of infinite sets.) Mathematicians today would say that there are no "self-evident truths." We can only *assume* the truth of certain statements. Then, granted that these axioms are so, we can prove other statements, called *theorems*. The "truth" of these theorems is not absolute, but relative. It depends on whether or not you accept the axioms upon which it is based.

Axioms are, of course, not chosen at random. They are selected because they are consistent with human experience.

An axiomatic approach to geometry is a formal one in which each statement is an axiom, a theorem, or a definition. While this book will not take such a formal approach, it will be interesting to look at a few statements typically accepted as axioms and see how these can be used to prove some theorems. The student may also come to understand what a "formal geometry" is and appreciate the reason for such an approach.

Our axioms concern the relationships between the undefined terms "point," "line," and "plane."

AXIOM 1   Two distinct points determine a unique line.

By "unique" we mean "one and only one." By "distinct" points we mean different points. Thus, this axiom says that given any two different points, there will be one and only one line containing them.

This axiom justifies our scheme of denoting a line by naming any two points on it, say p and q. Since by the axiom, p and q determine a unique line L, the symbol $\overleftrightarrow{pq}$ for the line is unambiguous. Moreover, if r and s are any other two points on L, then $\overleftrightarrow{pq} = \overleftrightarrow{rs}$; where here "$=$" denotes equality of sets, i.e., $\overleftrightarrow{pq}$ and $\overleftrightarrow{rs}$ are two names for the same set of points, namely the line L.

Points lying on one line will be called *collinear* and points lying in one plane will be called *coplanar*.

AXIOM 2   Three non-collinear points determine a unique plane.

Note that we did not say three *distinct* non-collinear points. Is the distinctness implied by the fact that the points are non-collinear? If you had three points, two of which coincided, could these three points be non-collinear?

Using this axiom we would be justified in describing a plane by naming any three points on it.

We can think of Axiom 2 as the "three-legged stool" axiom. A three-legged stool, even if its legs are not all the same length, will rest firmly on the ground, since the three points at the base of each leg determine a plane.

AXIOM 3   If two points lie in a plane, then the line containing them lies in the plane.

AXIOM 4   If two distinct planes intersect, then their intersection is a line.

AXIOM 5   A line contains at least two points.

Perhaps this axiom seems almost too trivial to be stated. Recall, however, that "line" is an undefined term, and we cannot make any assumptions about lines without stating these assumptions as axioms. This axiom is *not* the same as Axiom 1.

Now let us see how we can use these axioms to prove some simple theorems about the intersection of point sets in space.

THEOREM 1   Two distinct lines intersect in at most one point.

The proof of this theorem will be by contradiction (See Chapter 1). That is, we will assume the theorem to be false and show that this assumption leads us to a contradiction of one of our axioms.

Proof: Let L_1 and L_2 be two distinct lines and suppose that their intersection is *not* at most one point. Then, their intersection contains two points (or more), call them p and q. This means that we have two different lines L_1 and L_2 both containing p and q. But by Axiom 1 this is impossible since there is only one line containing p and q. ∎

Will the intersection of two distinct lines always be *exactly* one point?

THEOREM 2   If a line L intersects a plane Π, and L does not lie in the plane Π, then the intersection is a single point.

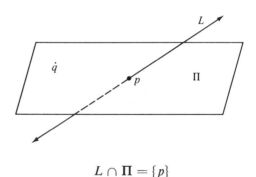

$$L \cap \Pi = \{p\}$$

Figure 2-6

Proof: We want to show that $L \cap \Pi$ is a single point. Again we will use a proof by contradiction. Suppose $L \cap \Pi$ contains more than one point, say points p and q. Then, p and q are both in L, and by Axiom 1 $\overleftrightarrow{pq} = L$. Since p and q are both in Π by Axiom 3, L must be in the plane Π, which contradicts our hypothesis. Hence, $L \cap \Pi$ contains at most one point. By hypothesis, $L \cap \Pi$ is not empty, hence the intersection contains *exactly* one point. ▮

THEOREM 3 Given a line L and a point p not on L, there is exactly one plane containing both of them.

Proof: Let q and r be two distinct points on L. By Axiom 5 we know that these two points exist. Then q and r are different from p, since p is not on L. By Axiom 2 the points p, q, and r determine a unique plane Π. Since q and r are in Π, by Axiom 3, L is in Π.

Thus, we have at least one plane, Π, containing p and L. If there were two such planes, then both of them would contain the three points p, q, and r, and this is impossible by Axiom 2. ▮

Note that we could not have proved this theorem without Axiom 5.

THEOREM 4 If two lines intersect then their union lies in exactly one plane.

The proof of this theorem is left as an exercise for the reader. (Problem 8)

Exercise 2.4

1. Are two distinct points always collinear? coplanar? Are three distinct points always collinear? coplanar? Explain.

2. We have an axiom that says two points are contained in exactly one line. Are two points contained in exactly one ray? Are two points contained in exactly one line segment? Draw diagrams to illustrate your answer.

3. How many planes can you find containing two distinct points? How many lines contain one point? Draw diagrams to illustrate your answer.

4. Suppose p, q, and r are three distinct points and all three points lie in two distinct planes Π_1 and Π_2. What can you say about the relationship between p, q, and r?

5. Let p_1, p_2, and p_3 be three non-collinear points. How many lines can you draw containing two of these points? Suppose you have four points in a plane no three of which are collinear. How many lines can you draw containing two of these points? Complete the following table:

Number of points in a plane no three of which are collinear	Number of lines determined
3	
4	
5	
6	
7	
.	
.	
.	
n	

6. How many lines are determined by four points that are *not* in the same plane? Draw a diagram to illustrate your answer.

7. Let p_1, p_2, p_3, and p_4 be four non-coplanar points. How many planes are determined by these four points? Suppose you have five points, no four of which are coplanar. How many planes do these five points determine? (Use Theorem 3.)

8. Prove Theorem 4 by giving reasons for each step below:
   Theorem 4. If two lines intersect, then their union lies in exactly one plane.
   Let L_1 and L_2 be two intersecting lines.
   (a) $L_1 \cap L_2$ is a single point p.
   (b) If q is a point on L_2 different from p, then there is a plane Π containing L_1 and q.
   (c) The plane Π contains L_2.
   (d) Π contains $L_1 \cup L_2$.
   (e) No other plane contains $L_1 \cup L_2$.

9. Let S be the set of all lines and Ⓡ be the relation "lies in the same plane as." Is this relation reflexive? Is it symmetric? Is it transitive? Illustrate your answers with a sketch.

5. Mathematical Models

We have so far a "geometry" consisting of three undefined terms and five axioms. Since the terms "point," "line," and "plane" are undefined, we can construct a mathematical model of our space by assigning meanings to these terms in such a way that the axioms will be true statements about these objects.

For example, the geometry we have developed in this chapter seems to be consistent with the physical world we live in. The elements "point," "line," and "plane" are taken to be our intuitive notions of these terms. This is not the only model for our geometry, however, and it will be instructive to look at some others.

Let us take our space to consist of four coins, a penny, a nickel, a dime and a quarter. A point in this space will be a coin. A line will be taken to be any subcollection of two distinct coins. Thus, for example, the set {penny, dime} will be a line in this space. Note that we have only four points and six lines in this rather poverty-stricken geometry. The lines are {penny, nickel}, {penny, dime}, {penny, quarter}, {nickel, dime}, {nickel, quarter}, and {dime, quarter}.

A plane will be any subset of three coins, for example, {penny, dime, quarter}. There are only four distinct planes. We can describe this geometry with the diagram in Figure 2-7.

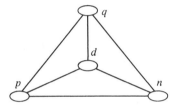

The four coin model

Figure 2-7

The student should verify that this model satisfies Axioms 1 through 5 as well as Theorems 1 through 4.

Now let us look at another model. Our space will be the set of points on the surface of a sphere. A "line" in this model will be a great circle. (A great circle is the intersection of the surface of the sphere with a plane through the center of the sphere.) This seems a reasonable interpretation of "line" on the surface of a sphere, since navigators know that the shortest distance between two points on our somewhat spherical earth is a great circle path. In this model we have only one plane—the surface of the sphere itself.

We will consider the endpoints of a diameter of the sphere to be a single point. For example, the North and South Poles will be one point. Thus, when we say "distinct" points, we will mean two different points which are not end-points of the same diameter. With this interpretation our five axioms are all satisfied. Since we have only one plane, some of the axioms are satisfied trivially. Thus, Axiom 4 says that *if* we have two distinct planes which intersect, then their intersection is a line. Since we do not have two distinct planes, we say this axiom is satisfied trivially.

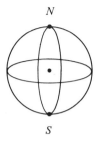

The model of the sphere

Figure 2-8

Note that Theorem 1 is true in the model, since although two great circles intersect in two points, these points are end-points of a diameter, hence are not considered to be different.

(Note that in this geometry *all* lines intersect; that is, there are no parallel lines! This does not contradict any of our axioms, however.)

As another model, consider the box in Figure 2-9. Let a "plane" be a side of the box, a "line" an edge, and a "point" a corner of the box. There are 6 planes, 12 lines, and 8 points in this geometry.

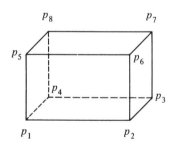

The box model

Figure 2-9

In this model, however, Axiom 1 is *not* satisfied, since we can find two distinct points (p_1 and p_6, for instance) that are not contained in any of the lines. Axiom 2 is not satisfied either, since there is no plane containing p_1, p_2, and p_8.

Axiom 3 *is* satisfied, however. We interpret Axiom 3 as follows: *If* two points lie in a plane, and *if* there is a line containing them, *then* this line lies in the plane. Axiom 3 does not assert the existence of a line containing any two points in the plane. This is done by Axiom 1. If Axiom 3 guaranteed the existence of this line, then we wouldn't need Axiom 1. The student should verify that Axioms 4 and 5 are satisfied in this model.

Exercise 2.5

1. How would you interpret "parallel lines" in the four coins model? Are there any parallel lines in this model? (See the definition of parallel lines in Section 1.)

2. Suppose that $\mathfrak{U}$ is a community consisting of just three citizens, A, B, and C. Suppose further that each pair of persons in $\mathfrak{U}$ forms a secret club excluding the third person. Take a "point" to be a person in $\mathfrak{U}$, a "line" to be a secret club, and "plane" to be the community $\mathfrak{U}$ itself. Restate Axioms 1 through 5 using "person" for point, "secret club" for line, and "community" for plane. Does this model satisfy the axioms?

3. Restate Theorems 1, 3, and 4 in terms of the model of problem 2. How would you interpret parallel lines in this model? Are there any parallel lines? How would you interpret "non-collinear points" in this model?

4. In the box model (Figure 2-9) Axioms 1 and 2 are not satisfied, but 3, 4, and 5 are. Which of Theorems 1 through 4 are satisfied by this model?

6. Intersections of Point Sets in Space

We can now use our axioms and theorems to investigate the relationship between two lines in space. There are two possibilities—either they intersect or they do not. If they intersect, then by Theorem 4, they lie in a plane, and by Theorem 1 their intersection is a single point. If they do not intersect, there are two cases to consider. Either they lie in the same plane, in which case we say they are parallel, or they are not in the same plane, and then we describe them as *skew* lines.

Let us summarize. Given two distinct lines L_1 and L_2 in space, three possible cases arise.

Case 1 $L_1 \cap L_2$ is not empty.

Then, L_1 and L_2 must lie in a plane and their intersection is a single point.

Case 2 $L_1 \cap L_2 = \varnothing$ and L_1 and L_2 are in the same plane. Then, L_1 and L_2 are parallel.

Case 3 $L_1 \cap L_2 = \emptyset$ and L_1 and L_2 are not in the same plane. Then, L_1 and L_2 are skew.

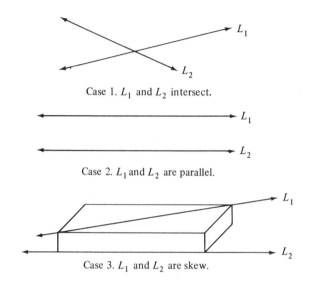

Case 1. L_1 and L_2 intersect.

Case 2. L_1 and L_2 are parallel.

Case 3. L_1 and L_2 are skew.

Figure 2-10

If we look at the relationship between two planes in space, only two cases arise. Either their intersection is empty, or it is not. If their intersection is empty, then the two planes are said to be parallel. For example, the two opposite sides of a room can be thought of as portions of parallel planes. If two planes intersect, then their intersection must be a line. (Axiom 4) See Figure 2-11.

Let Π_1 and Π_2 be two distinct planes.

Case 1 $\Pi_1 \cap \Pi_2 = \emptyset$

The two planes are parallel.

Case 2 $\Pi_1 \cap \Pi_2 \neq \emptyset$

Their intersection is a line.

There are three possible ways a line L and a plane Π in space may be related. The line might lie in the plane, the line might intersect the plane in exactly one point (Theorem 2), or their intersection might be empty.

Case 1 The line L lies in the plane Π. $L \cap \Pi = L$

Case 1. Π_1 and Π_2 are parallel.

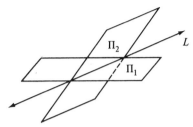

Case 2. Π_1 and Π_2 intersect in the line L.

Figure 2-11

Case 1. The line L lies in the plane Π.

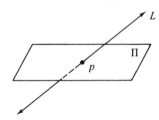

Case 2. L intersects Π in exactly one point.

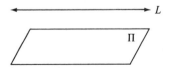

Case 3. L and Π do not intersect.

Figure 2-12

Case 2   The line L intersects the plane in exactly one point, p.

$$L \cap \Pi = \{p\}$$

Case 3 $L \cap \Pi = \varnothing$

Exercise 2.6

1. Given two distinct rays in a plane list all possible forms that their intersection might take. Sketch each.

2. Given two distinct line segments in a plane list all possible forms that their intersection might take. Sketch each.

3. Let L_1 and L_2 be two parallel lines. Into how many disjoint sets does the set $L_1 \cup L_2$ separate the plane? Describe these sets in terms of half-planes, where Π_1 and Σ_1 denote the half-planes determined by L_1, and Π_2 and Σ_2 denote the half-planes determined by L_2.

4.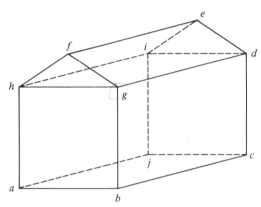

   The figure above suggests a number of lines and planes. The lines may be described by pairs of points and the planes may be described by naming three points. Name

   (a) a pair of parallel planes.
   (b) a pair of planes whose intersection is a line.
   (c) three planes whose intersection is a single point.
   (d) three planes whose intersection is a line.
   (e) a line and a plane whose intersection is empty.
   (f) a line and a plane whose intersection is a point.
   (g) a line and a plane whose intersection is a line.
   (h) a pair of parallel lines.
   (i) a pair of skew lines.
   (j) three lines that intersect in a single point.
   (k) four planes whose intersection is exactly one point.

5. Suppose L_1 and Π are a line and a plane in space whose intersection is empty. Let L_2 be a line in the plane Π. What can you say about the relationship between L_1 and L_2? Sketch.

6. Suppose that you have three distinct lines all lying in the same plane. Find all the possible ways in which the three lines can be related. Sketch each. (4 cases)

7. Suppose that you have three distinct planes in space. Find all the possible ways these three planes can be related. Sketch. (5 cases)

8. Suppose that you have three distinct lines in space (not necessarily in the same plane). Find all the possible ways these three lines might be related. Sketch. (At least 9 cases)

9. Let S be the set of lines in a plane and let ⓇＲ be the relation "intersects." We say L_1 intersects L_2 if their intersection is not empty. Is this relation reflexive? Is it symmetric? Is it transitive?

7. Unions of Point Sets in Space

Angles

Let us consider the union of two distinct rays with a common end-point. If they are opposite rays, then their union is a line. Suppose they are not opposite rays.

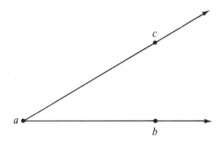

An angle is the union of two rays.

Figure 2-13

We will define an *angle* to be the union of two rays with a common end-point which do not lie on the same line. Thus, an angle is a set of points in a plane. Note that we do not consider a straight line to be an angle.

This definition may seem a little strange at first. You have perhaps thought vaguely of an angle as the amount of the opening between two rays or something of that sort. Such a notion is more an idea of the measure of an angle than of the angle itself. We want our definition of an angle to be independent of its measure. Later, we will discuss the measure of an angle, but at this stage an angle to us will be simply a certain set of points in space.

If a is the common end-point of the two rays $\overrightarrow{ab}$ and $\overrightarrow{ac}$, then a is called the *vertex* of the angle. We will designate the angle by the symbol $\angle\,cab$ (or perhaps $\angle\,bac$. The order of the letters is not important as long as the letter naming the vertex is listed second.) Sometimes, if there is no chance of confusion, we may call this angle $\angle\,a$. The rays $\overrightarrow{ab}$ and $\overrightarrow{ac}$ are called the *sides* of the angle.

$$\angle\,cab = \overrightarrow{ab} \cup \overrightarrow{ac}$$

An angle divides the plane into three disjoint point sets. One is, of course, the angle itself; the other two will be called the *interior* and the *exterior* of the angle, respectively. It is fairly obvious from the figure which set we want to call the interior of $\angle\,cab$, but we need to describe this set so that our definition will not depend on a picture. Since $\angle\,cab$

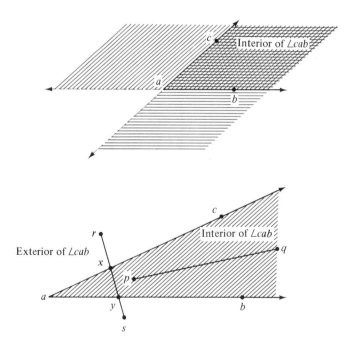

An angle divides the plane into three disjoint point sets, the angle itself, its interior and its exterior.

Figure 2-14

is the union of two rays, $\overrightarrow{ab}$ and $\overrightarrow{ac}$, we can use the two lines $\overleftrightarrow{ab}$ and $\overleftrightarrow{ac}$ containing these rays to define the interior of the angle. We have already seen that $\overleftrightarrow{ab}$ and $\overleftrightarrow{ac}$ separate the plane into two half-planes. We can accurately describe the interior of $\angle cab$ as the intersection of two of these half-planes. Which two shall it be? Line $\overleftrightarrow{ab}$ separates the plane into two half-planes one of which contains the point c. Similarly, $\overleftrightarrow{ac}$ separates the plane into two half-planes one of which contains the point b. The intersection of these two half-planes will be called the interior of $\angle cab$. The third set remaining will be called the exterior of $\angle cab$. Briefly then, the interior of $\angle cab$ is the intersection of the half-plane on the b side of $\overleftrightarrow{ac}$ and the half-plane on the c side of $\overleftrightarrow{ab}$.

There is still another way of distinguishing the interior of an angle from its exterior. Given any two points p and q in the interior of an angle, the line segment joining them contains no points of the angle itself.

On the other hand, it is possible to find points r and s in the exterior of the angle so that the line segment $\overline{rs}$ does contain points of the angle (points x and y in Figure 2-14).

Vertical and Adjacent Angles

When two lines intersect, a number of angles are formed. (How many?)

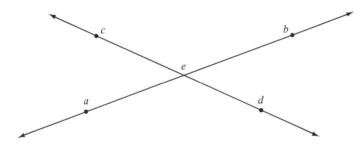

$\angle bed$ and $\angle cea$ are vertical angles, as are $\angle ceb$ and $\angle aed$.

Figure 2-15

Certain pairs of these angles have a special relationship. In Figure 2-15 $\angle bed$ and $\angle cea$ are called *vertical* angles, as are $\angle ceb$ and $\angle aed$. Two angles are called vertical angles if their union is two lines.

If two angles have the same vertex and their intersection is a ray, then they are called *adjacent* angles.

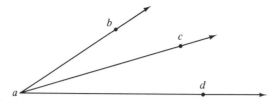

(a) ∠ *bac* and ∠ *cad* are adjacent angles.

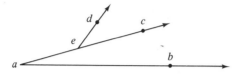

(b) ∠ *dec* and ∠ *cab* are not adjacent angles.

Figure 2-16

In Figure 2-16(a), ∠ *bac* and ∠ *cad* are adjacent since they share the same vertex *a* and their intersection is the ray $\overrightarrow{ac}$. In Figure 2-16(b) ∠ *cab* and ∠ *dec* are not considered to be adjacent. (Why?)

Polygons

If we take three points $p_1, p_2,$ and p_3 in space, not all on the same line (non-collinear), then these points determine three line segments, $\overline{p_1p_2}$, $\overline{p_2p_3}$, and $\overline{p_3p_1}$.

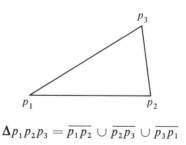

$$\Delta p_1p_2p_3 = \overline{p_1p_2} \cup \overline{p_2p_3} \cup \overline{p_3p_1}$$

Figure 2-17

The union of these three line segments is called a *triangle*. The points $p_1, p_2,$ and p_3 are called the *vertices* of the triangle and the line segments are called its *sides*.

In the same way, the union of the four line segments $\overline{p_1p_2}$, $\overline{p_2p_3}$, $\overline{p_3p_4}$, and $\overline{p_4p_1}$, in Figure 2-18 is called a *quadrilateral*.

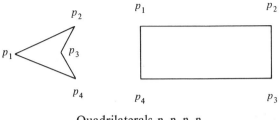

Quadrilaterals $p_1p_2p_3p_4$

Figure 2-18

Let us be more specific about the arrangement of these line segments. For example, we would not want to consider Figure 2-19 a quadrilateral although it is the union of the line segments $\overline{p_1p_2}$, $\overline{p_2p_3}$, $\overline{p_3p_4}$, and $\overline{p_4p_1}$.

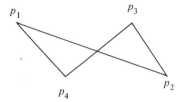

$p_1p_2p_3p_4$ is not a quadrilateral.

Figure 2-19

We can define a quadrilateral in this way: Let p_1, p_2, p_3, and p_4 be four points all in the same plane, no three of which are collinear. Then, if the line segments $\overline{p_1p_2}, \overline{p_2p_3}, \overline{p_3p_4}, \overline{p_4p_1}$ intersect only at their end-points, their union will be called a *quadrilateral*. What kind of figure would we get if three of the points were collinear?

If n distinct points $(n \geq 3)$ $p_1, p_2, \ldots, p_n$ all in the same plane are arranged in such a way that no three consecutive points are collinear and the line segments $\overline{p_1p_2}, \overline{p_2p_3}, \ldots, \overline{p_np_1}$ intersect only at their end-points, then the union of these n segments is called a *polygon*. Thus, triangles and quadrilaterals are polygons which are the unions of three and four line segments, respectively. There are also special names for polygons with more than four sides. A five sided polygon is called a *pentagon;* one with six sides a *hexagon;* one with seven a *heptagon;* eight, an *octagon;* nine, a *nonagon;* and ten, a *decagon.* The end-points are called the *vertices* of the polygon and the line segments are called its *sides.* It will be convenient to describe a polygon by naming its consecutive vertices.

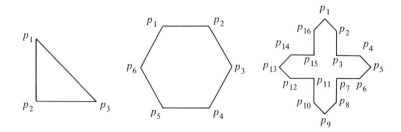

Polygons

Figure 2-20

Thus, we may speak of the quadrilateral $p_1p_2p_3p_4$, or of the triangle *abc*. We will also use the symbol $\triangle abc$ to denote the triangle *abc*.

A line segment joining non-consecutive vertices of a polygon is called a *diagonal*. Thus, in Figure 2-21, $\overline{p_1p_3}$, $\overline{p_1p_4}$, and $\overline{p_1p_5}$ are diagonals. These are, of course, not all of the diagonals. Line segments $\overline{p_2p_4}$, $\overline{p_2p_5}$, and $\overline{p_2p_6}$ will also be diagonals. How many diagonals does a hexagon have?

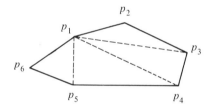

$\overline{p_1p_3}$, $\overline{p_1p_4}$ and $\overline{p_1p_5}$ are diagonals of the polygon $p_1p_2p_3p_4p_5p_6$.

Figure 2-21

Any polygon has a number of angles associated with it, one for each vertex. Consider the triangle *abc* in Figure 2-22.

This triangle determines three angles, $\angle cab$, $\angle abc$, and $\angle acb$. Since there is no danger of confusion in this case we can refer to them as $\angle a$, $\angle b$, and $\angle c$. Note that these angles are *not* subsets of the triangle *abc*. For example, $\angle a$ is the union of two rays, $\overrightarrow{ab}$ and $\overrightarrow{ac}$. Clearly, these rays contain points which are not in triangle *abc*. However, we will in the future refer to the angles of a triangle, keeping in mind that these are the angles associated with the triangle and are not subsets of it.

In the same way we will speak of the four angles of a quadrilateral and the *n* angles of a polygon with *n* vertices.

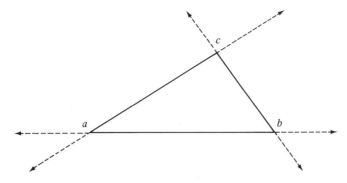

$\angle a$, $\angle b$ and $\angle c$ are associated with $\triangle abc$, but they are not subsets of $\triangle abc$.

Figure 2-22

Exercise 2.7

1. Draw in one plane two different angles whose intersection is:

   (a) empty. (b) one point.
   (c) two points. (d) three points.
   (e) four points. (f) a line segment.
   (g) a ray.

2. Exactly one of the three disjoint subsets determined by an angle can contain a line. Which subset is this? Using this idea, formulate a definition for the exterior and interior of an angle.

3. In the following figure, let I denote the interior of $\angle abc$, E its exterior and Π the plane. Tell whether each of the following statements is true or false. If a statement is false, rewrite it to make it true.

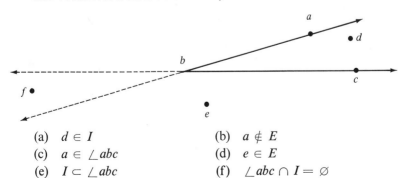

   (a) $d \in I$ (b) $a \notin E$
   (c) $a \in \angle abc$ (d) $e \in E$
   (e) $I \subset \angle abc$ (f) $\angle abc \cap I = \varnothing$

(g) $I \subset \Pi$ (h) $I \cap E = \angle abc$

(i) $f \in I$ (j) $\angle abc \subset \Pi$

4. Let R_1, R_2, and R_3 be distinct non-opposite rays having the same end-point. How many different angles are formed by their union? Suppose you have four distinct non-opposite rays having the same end-point. How many different angles are formed by their union? Complete the following table:

Number of distinct, non-opposite rays having the same end-point	Number of angles formed
2	
3	
4	
5	
.	
.	
.	
n	

5. A collection of lines is said to be *concurrent* if all the lines intersect in the same point. Given two concurrent lines, how many pairs of vertical angles can be formed? How many pairs of vertical angles are formed by three concurrent lines? Complete the following table:

Number of concurrent lines	Number of pairs of vertical angles formed
2	
3	
4	
5	
.	
.	
.	
n	

6. Let S be the set of angles and let the relation ⓡ be "is adjacent to." Is this relation reflexive? Is it symmetric? Is it transitive? Explain.

7. Draw a triangle and a line segment so that the intersection of the two point sets consists of

(a) one point.   (b) two points.

(c)   a line segment.   (d)   the empty set.

8.   Draw two triangles whose intersection is:

(a)   empty.   (b)   one point.
(c)   two points.   (d)   three points.
(e)   four points.   (f)   five points.
(g)   six points.   (h)   a line segment.

9.

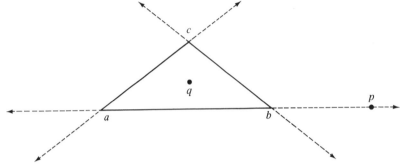

Use the above figure to tell whether each of the following statements is true or false. If a statement is false, rewrite it making it true.

(a) $q \in \triangle abc$ (b) $a \in \triangle abc$
(c) $\angle a \cup \angle b \cup \angle c = \triangle abc$ (d) $p \in \angle a$
(e) $p \in \angle b$ (f) $p \in \triangle abc$
(g) $\angle a \cup \angle b \cup \angle c = \overleftrightarrow{ab} \cup \overleftrightarrow{ac} \cup \overleftrightarrow{bc}$

10.   How many diagonals can be drawn from *one* vertex in a quadrilateral? In a pentagon? Complete the following table:

Number of sides	Number of diagonals from one vertex
4	
5	
6	
7	
.	
.	
.	
n	

11.   How many distinct diagonals are there in a quadrilateral? In a pentagon? Complete the following table:

Number of sides	Number of distinct diagonals
4	
5	
6	
7	
.	
.	
.	
n	

12. Does the figure below fit the definition of a polygon? Why or why not?

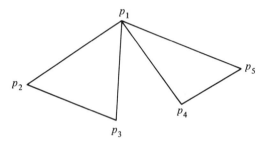

8. Simple Closed Curves

A polygon is an example of a special type of geometric figure called a *simple closed curve* (scc). In order to explain what a scc is, let us take these three words one at a time.

Intuitively, a curve is any geometric figure that can be drawn without lifting the pencil from the paper. By this we mean to suggest that the figure is continuous—there are no breaks in it.

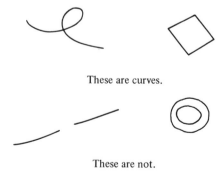

These are curves.

These are not.

Figure 2-23

Although it may seem strange to use the term this way, a curve may be a line segment or a polygon. Here we are using the term "curve" in a technical mathematical sense, not as an adjective to describe something "curvy." A curve may be infinite in extent. For example, lines, rays, and angles are curves. Although we cannot physically draw a line in one continuous movement of the pencil, our intuitive idea of a line is of a figure with no breaks in it.

A curve is *closed* if the starting point and the end-point coincide. A line is not a closed curve because it has neither starting point nor end-point.

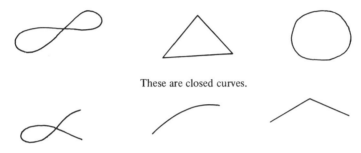

These are closed curves.

These curves are not closed.

Figure 2-24

In mathematics, the opposite of "closed" is not necessarily "open." A door is open if it is not closed, but curves are not doors, and we will not use the adjective "open" to describe a curve. If the starting point and the end-point of a curve are not the same, we simply say the curve is not closed.

A closed curve is *simple* if it can be drawn without tracing the same point twice. (Except, of course, for the starting point.) A simple closed curve

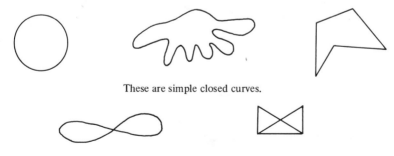

These are simple closed curves.

These are closed curves which are not simple.

Figure 2-25

does not cross itself. The adjective "simple" will be applied only to closed curves. If a curve is not closed, then it cannot be classified as either simple or not simple.

Many of the curves that we study in geometry are simple closed curves. Polygons and circles are scc's.

A simple closed curve divides the plane into three disjoint sets of points: the curve itself, its interior, and its exterior. This is not a definition, however, for there are curves which divide the plane into three disjoint point sets but which are *not* scc's. An angle is one example.

The interior of a scc has a property which the interior of an angle does not have. It is a *bounded* set of points. A set of points in a plane is *bounded* if we can draw a circle large enough to enclose all of the points of the set.

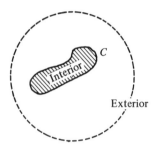

A simple closed curve *C*. The interior is bounded. The exterior is not bounded.

Figure 2-26

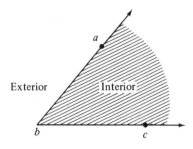

∠ *abc* is not a scc. Note that neither its interior nor its exterior is bounded.

Figure 2-27

We can picture the relation between the various geometric figures we have discussed in Figure 2-28.

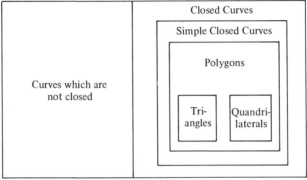

Curves in a plane

Figure 2-28

A *region* in the plane is the union of a simple closed curve and its interior. The triangular region in Figure 2-29 is the union of $\triangle abc$ and its interior.

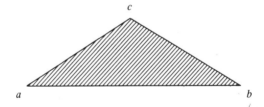

A region in the plane is the union of a scc and its interior.

Figure 2-29

9. Convex Sets

A set A is said to be *convex* if for every pair of points p and q in A, the entire line segment $\overline{pq}$ is contained in the set A. For example, the triangular region in Figure 2-30 is a convex set. No matter what pair of points we pick in this region, the line segment joining them lies wholly in the region. We have illustrated this with line segments $\overline{p_1q_1}$, $\overline{p_2q_2}$, and $\overline{p_3q_3}$.

On the other hand, none of the regions in Figure 2-31 is convex. In each case it is possible to find a pair of points, p and q, so that the line segment $\overline{pq}$ does not lie in the region.

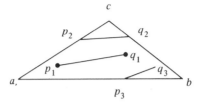

A triangular region is a convex set.

Figure 2-30

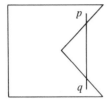 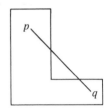

These regions are not convex.

Figure 2-31

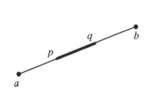 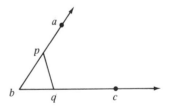

$\overline{ab}$ is a convex set. $\angle abc$ is not a convex set.

Figure 2-32

A convex set does not have to be a region. For example, a line is a convex set, as is a line segment or a ray. An angle, however, is not a convex set.

A convex set can be very large. The whole plane is convex, and so is a half-plane. If a polygonal region is convex, then every diagonal of the polygon lies in its interior.

Exercise 2.9

1. Classify the following curves as closed or not closed. Classify the closed curves as simple or not simple.

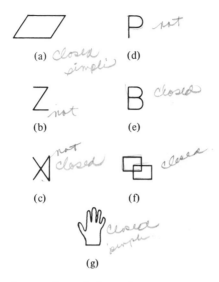

(a) *closed simple*   (d) *not*

(b) *not*

(e) *closed*

(c) *not closed*

(f) *closed*

(g) *closed simple*

2. Which of the following are bounded sets?

(a) an angle *no*   (b) the interior of an angle *no*

(c) the exterior of an angle *no*   (d) a triangle *yes*

(e) a single point *yes*   (f) a line *no*

(g) a ray *no*   (h) a line segment *yes*

(i) a polygon *yes*   (j) a polygonal region *yes*

3. Tell whether or not the following sets are convex.

(a) an angle *no*   (b) the interior of an angle *yes*

(c) the exterior of an angle *no*   (d) the interior of a triangle *yes*

(e) the exterior of a triangle *no*   (f) a triangle *no*

(g) a single point *yes*   (h) the empty set *yes*

4. If a quadrilateral is convex, its diagonals intersect. Is this true if the quadrilateral is not convex? Draw a figure to illustrate your answer. *not necessary.*

5. Is the intersection of two convex sets a convex set? Is the union of two convex sets a convex set? Draw a number of different figures to illustrate your answers.

6. State whether the following statements are true or false. If the statement is false, change it to make it true.

(a) All triangles are polygons. *true*

(b) Some polygons are triangles. *true*

(c) All polygons are triangles. *see no*

(d) A quadrilateral is a simple closed curve. *yes*

(e)   A simple closed curve is a polygon.

(f)   No triangles are quadrilaterals.

7.   Draw a triangle *abc* and label points *x* and *y* in its interior.

(a)   What is $\overline{xy} \cap \triangle abc$?

(b)   What can you say about the relation between the set of points on $\overline{xy}$ and the set of points in the interior of $\triangle abc$?

(c)   Draw points *p* and *q* in the exterior of $\triangle abc$ so that $\overline{pq} \cap \triangle abc = \varnothing$.

(d)   Draw points *r* and *s* in the exterior of $\triangle abc$ so that $\overline{rs} \cap \triangle abc \neq \varnothing$.

(e)   Can you find two points *u* and *v* in the interior of $\triangle abc$ such that $\overline{uv} \cap \triangle abc \neq \varnothing$?

(f)   Use these results to formulate a definition of the interior of $\triangle abc$; of the exterior of $\triangle abc$.

New Terms Found in This Chapter

Term	Section	Term	Section
parallel lines	1	interior of an angle	7
intersect	2	exterior of an angle	7
half-line	3	vertical angles	7
ray	3	adjacent angles	7
end-point of a ray	3	triangle	7
opposite rays	3	quadrilateral	7
line segment	3	polygon	7
half-planes	3	diagonal	7
axiom	4	concurrent lines	Exercise 2.7
collinear	4	curve	8
coplanar	4	closed curve	8
skew lines	6	simple closed curve	8
angle	7	bounded set	8
vertex of an angle	7	region in the plane	8
sides of an angle	7	convex set	9

3: Congruence and Measure

1. Congruence

We say intuitively that two geometric figures are congruent if one can be moved or transferred so that it coincides with the other, point for point. In other words, congruent figures have the same size and shape. We assume that in moving a geometric figure we do not change any of its properties, except of course, its location in space. These figures are "rigid" when moved; that is, they do not bend or stretch out of shape. Thus, if we move one triangle so that it coincides with another, the transfer does not alter the size of any of the angles nor the lengths of any of the sides. We can see that the notion of congruence and the notion of measure are very closely related. In what follows we will study the idea of measure of line segments and of angles, and define congruence of these figures in terms of their measure.

2. Measure and Congruence of Line Segments

Measure

To measure line segments we use units of length. Early units of length were parts of the body. A *cubit* was the length of a man's arm, from the elbow to the end of the middle finger. The distance from the end of the thumb to the end of the little finger when the hand was spread out was the *span*. Two spans made a cubit. The *fathom* was the distance from finger tip to finger tip when the arms were outstretched. It was about six feet and is still used today in measuring ocean depths. A *pace* was two steps and 1,000 paces were a *mile*, from the Latin *mille* meaning a thousand.

The story goes that the *yard* was set by Henry I as the distance from the tip of his nose to the end of the thumb of his outstretched arm. The *inch* was established to be the length of three barley-corns laid end to end.

These units were not standardized, to say the least. A more modern unit of measure is the *meter*, set in 1791 as one ten-millionth of the distance from the North Pole to the Equator. A meter is 39.17 inches, a little longer than a yard.

Now let us consider a line *L*, and on it choose an arbitrary point which we shall label "0." We then choose another point on the line and label it "1."

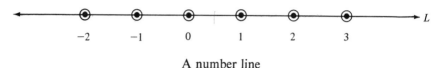

$$-2 \qquad -1 \qquad 0 \qquad 1 \qquad 2 \qquad 3$$

A number line

Figure 3-1

With this modest beginning we can construct a *number line*—a line *L* such that every point on *L* is labeled with exactly one real number, and conversely, every real number is a label for exactly one point on *L*. Thus, we can set up a one-to-one correspondence between the set of real numbers and the points on the line *L*.

For example, to find those points corresponding to the positive integers, we will call the distance from the point 0 to the point 1 our unit. To the number 2 we assign the point one unit to the right of 1, and in the same way, we assign to the point one unit to the left of 0 the integer -1. This correspondence between the real numbers and the points of *L* is called a *coordinate system*. The correspondence is defined in such a way that the distance between any two consecutive integers is the same. The real number assigned to the point *p* is called its *coordinate*. Because of this one-to-one correspondence, we will often identify a point by its coordinate and refer to "the point 2" when we mean "the point with coordinate 2."

We will assume that every line has a coordinate system; in fact, it is not hard to see that every line can have many coordinate systems. We will call this assumption the "ruler axiom," for essentially it states that given any line we can construct from it a "ruler" with which we can measure lengths. If this axiom seems too obvious to deserve comment, consider for a moment the geometry of the sphere discussed in Chapter 2. Here "lines" are great circles and are finite in length, hence these "lines" do not have a coordinate system and the ruler axiom does not hold in this model.

If we want to measure a line segment $\overline{pq}$, we can use one of the coordinate systems of the line $\overleftrightarrow{pq}$ to do so. The symbol we will use for the measure of $\overline{pq}$ is *pq*. At this point it will be helpful to review the different symbols we have used involving the points *p* and *q*. Recall that $\overleftrightarrow{pq}$ means the *line* containing the points p and q; $\overrightarrow{pq}$ means the *ray* with end-point p containing the point q; $\overline{pq}$ means the *line segment* with end-points p and q; and finally pq means the measure of the line segment $\overline{pq}$.

In order to compare the measures of different line segments, we will use the same unit length for all our coordinate systems. We will not specify what kind of a unit this is, but will simply refer to it as our "unit of length."

Now suppose we have a line L to which we have assigned some coordinate system. Then given any two points p and q on the line, no matter how

If x_1 is the coordinate of p and x_2 is the coordinate of q then the measure of $\overline{pq}$ is the number $|x_1 - x_2|$.

Figure 3-2

they are situated, we can define the measure of the line segment $\overline{pq}$ as the distance between them. If x_1 is the real number assigned to p and x_2 is the real number assigned to q, this distance will be denoted by $|x_1 - x_2|$.

Thus
$$pq = |x_1 - x_2|$$

where
$$|x_1 - x_2| = x_1 - x_2 \quad \text{if} \quad x_1 > x_2$$
and
$$|x_1 - x_2| = x_2 - x_1 \quad \text{if} \quad x_2 > x_1$$

From this we can see that the measure of $\overline{pq}$ is the same as the measure of $\overline{qp}$. This would seem to be a highly desirable result.

For example, if p has coordinate 5 and q has coordinate 2, then

$$pq = |5 - 2| = qp = |2 - 5| = 3$$

If p has coordinate 2 and q has coordinate -1, then

$$pq = |2 - (-1)| = qp = |(-1) - 2| = 3$$

Properties of Measure of Line Segments

From the way in which we have described the measure of a line segment, we can see that this measure has the following properties:

1. The measure of a line segment is always a positive real number.
2. If $\overline{pq} \subset \overline{ab}$, then $pq \leq ab$.

This says that if the line segment $\overline{pq}$ is contained in the line segment $\overline{ab}$, then the measure of $\overline{pq}$ is less than (or perhaps equal to) the measure of $\overline{ab}$.

3. If $\overline{pq}$ is a line segment and if r is some point between p and q, then $pr + rq = pq$. This is known as the *additive* property of the measure of line segments.

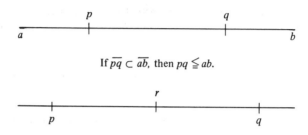

If $\overline{pq} \subset \overline{ab}$, then $pq \leqq ab$.

If r is between p and q, then $pr + rq = pq$.

Figure 3-3

Midpoint of a Line Segment

If $\overline{pq}$ is a line segment, then there is exactly one point r on $\overline{pq}$ such that $pr = rq$. This point r is called the *midpoint* of the segment $\overline{pq}$. Since by the additive property of measure, $pr + rq = pq$, we have $2pr = pq$ or $pr = \frac{1}{2}pq$. That is, the measure of the segment $\overline{pr}$ is one-half the measure of the segment $\overline{pq}$. In the same way $rq = \frac{1}{2}pq$. If r is the midpoint of $\overline{pq}$, then we sometimes say that r *bisects* the segment $\overline{pq}$.

If we know the coordinates of two of the points, p, q, and r, then we can find the coordinate of the third. Thus, if p has coordinate 2 and q has coordinate 3, then the midpoint r has coordinate $2\frac{1}{2}$. If the midpoint has coordinate -2, and p has coordinate -13, then q has coordinate 9. (Figure 3-4) Note that if p and q have coordinates x_1 and x_2 respectively, then the coordinate of the midpoint is $(x_1 + x_2)/2$.

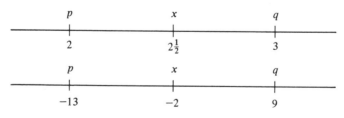

x is the midpoint of $\overline{pq}$.

Figure 3-4

Congruence of Line Segments

Now that we have described the measure of line segments we can define a congruence relation between line segments. Two line segments are said to be *congruent* if they have the same measure. The symbol we will use for this relation is $\cong$. Thus if $\overline{pq}$ and $\overline{rs}$ are congruent (i.e., $pq = rs$), then we will write $\overline{pq} \cong \overline{rs}$. This is not the same thing as saying $\overline{pq} = \overline{rs}$, since $\overline{pq}$ and $\overline{rs}$ are sets of points, and equality between sets means that every point in the first set is in the second, and every point in the second set is in the first. In particular, if $\overline{pq} = \overline{rs}$, then $\overline{pq}$ and $\overline{rs}$ are two names for the same line segment.

If $\overline{pq} = \overline{rs}$, then $\overline{pq}$ and $\overline{rs}$ obviously have the same measure, hence $\overline{pq} \cong \overline{rs}$. However, two line segments can be congruent without being equal.

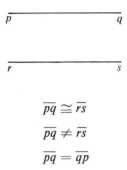

$$\overline{pq} \cong \overline{rs}$$
$$\overline{pq} \neq \overline{rs}$$
$$\overline{pq} = \overline{qp}$$

(The symbol $\neq$ means "is not equal to.")

Figure 3-5

Exercise 3.2

1. Find pq if the following real numbers are assigned to p and to q:

p	q	pq
-1	$1\frac{1}{2}$	
0	-7	
-10	-3	
$2\frac{1}{2}$	14	
-2	14	

2. Let a, b, and c be three distinct points in a plane.
   (a) If $ab + bc = ac$, what can you say about the relationship between the three points a, b, and c?

(b) Could a, b, and c be situated so that

$$ab + bc > ac? \qquad ab + bc < ac?$$

Explain.

3. A relation Ⓡ on a set S is said to be an equivalence relation if it satisfies the following properties: (Let a, b, and c represent elements of S.)

  (i) The Reflexive Property. a Ⓡ a for all $a \in S$.
  (ii) The Symmetric Property. Whenever a Ⓡ b, then b Ⓡ a.
  (iii) The Transitive Property. Whenever a Ⓡ b, and b Ⓡ c, then a Ⓡ c.

  Show that if S is the set of all line segments and Ⓡ is the relation "is congruent to," then Ⓡ is an equivalence relation.

4. Let a and b be points on a number line and let c be the midpoint of $\overline{ab}$. Complete the following table. Sketch.

a	b	c
0	1	
-1		2
$\frac{1}{2}$	-1	
	1	3
1	-2	

5. Let a, b, c, d, and e be points on a number line to which are assigned the coordinates -1, $-1/3$, 0, $1\frac{1}{2}$, and 5, respectively. Find the coordinates of the midpoints of the segments $\overline{ab}$, $\overline{ac}$, $\overline{ad}$, $\overline{ae}$, $\overline{bc}$, $\overline{bd}$, $\overline{be}$, $\overline{cd}$, $\overline{ce}$, $\overline{de}$.

6. If x_1, x_2, x_3, x_4, x_5, x_6, x_7, x_8, x_9, and x_{10} are the midpoints of each of these segments, find ax_1, ax_2, ax_3, ax_4, bx_5, bx_6, bx_7, cx_8, cx_9, dx_{10}.

7. Let p, q, and x be points such that $px = xq$. Is x necessarily the midpoint of $\overline{pq}$? Why or why not? How many points x satisfy the relation $px = xq$? Draw some of these points. What do you notice about the relation between these points?

8. If $pq = rs$, which of the following statements are true?
  (a) $\overline{pq} \cong \overline{rs}$
  (b) $\overline{pq} \cong \overline{sr}$
  (c) $p = r$ and $q = s$
  (d) $pq + rs = 2pq$
  (e) $pq - rs = 0$
  (f) $\overline{pq} \cap \overline{rs} = \varnothing$
  (g) $\overline{pq} = \overline{rs}$

9. Let p be a point with the coordinate given below. Find the coordinates

of the points q_1 and q_2 so that the segment $\overline{pq_1}$ and $\overline{pq_2}$ will have the given length.

coordinate of p	length
0	2
2	3
$1\frac{1}{2}$	$\frac{1}{4}$
$\sqrt{2}$	2
-1	$2\frac{1}{2}$
-7	4

3. The Circle

We are all familiar with the pleasantly symmetric appearance of the circle. The Greeks considered the circle to be the perfect figure. Because they considered the heavens also to be perfect, they concluded that the planets must move in circular orbits. Their logic was faulty, but we can understand why the orderly picture of heavenly bodies moving in circular orbits appealed to them.

Formally, we describe a circle as follows: If p is a point, and r is some positive real number, then the circle with center at p and radius r is the set of all points q in a plane whose distance from p is equal to r.

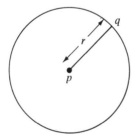

A circle with center at p and radius r. The line segment $\overline{pq}$ is also called a radius and its measure is the number r.

Figure 3-6

If q is any point on the circle, then the line segment $\overline{pq}$ is called a *radius* of the circle. Note that we have used the word "radius" in two ways. We speak of the radius as a line segment, or we refer to the real number r as the radius. Thus we might say "the circle with radius 2," or on the other hand, we might speak of congruence of two radii. A line segment through the center of a circle, with end points on the circle is called a *diameter*. Its measure is also referred to as the diameter of the circle. Referring to the diameter and radius as point sets, we might say that the diameter is the union of two radii which lie on the same line. Or we could say that the diameter is twice the radius, in which case we are clearly thinking of diameter and radius as numbers.

A circle is a simple closed curve. It divides the plane into three disjoint point sets: the circle itself, its interior, and its exterior. A circular region, the union of a circle and its interior, is a convex set.

It is easy to tell whether a given point q is on the circle, in the interior of the circle, or in its exterior by measuring the distance from q to the center of the circle, p. If $pq = r$, then q is on the circle. If $pq < r$, then q is in the interior of the circle, and if $pq > r$, q is in the exterior of the circle. (Figure 3-7)

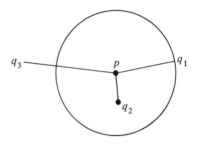

q_1 is on the circle, $pq_1 = r$; q_2 is in the interior of the circle, $pq_2 < r$; q_3 is in the exterior of the circle, $pq_3 > r$.

Figure 3-7

The interior of the circle with center at p and radius r is the set of all points whose distance from p is less than r, and the exterior is the set of all points whose distance from p is greater than r.

Exercise 3.3

1. Given a circle and a line in the same plane, there are three possibilities for their intersection. Describe these three cases and sketch.

2. Let C_1 and C_2 denote circles with centers at a and b and radii r and s, respectively. Suppose $a \neq b$ and let $t = ab$. For each of the following cases, draw a picture showing the relationship between the circles C_1 and C_2.

(a) $r + s < t$ (b) $r + s > t$
(c) $r + s = t$ (d) $r + t = s$
(e) $r + t > s$ (f) $r + t < s$
(g) $r = t > s$ (h) $r = t < s$

3. In the definition of a circle, the phrase "in a plane" appears. If this phrase is left out, what kind of geometric figure does the definition describe?

4. Measure and Congruence
of Angles

Measure of Angles

To measure line segments we set up a coordinate system on a line. Our "ruler," being a line, is infinite in length. Any positive real number, no matter how large, can be thought of as the length of some line segment.

The "ruler" that we use to measure angles, however, will not be infinite in length. To measure angles we use a half-circle (a semi-circle).

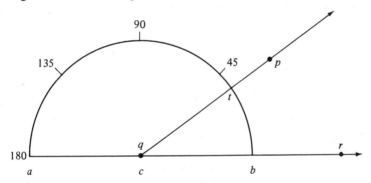

Using a semi-circle to measure angles.

Figure 3-8

Consider the semi-circle in Figure 3-8 with center at c. We will assume that the circular arc from a to b can be divided up into n equal parts for any positive integer n. We might call this our "Protractor Axiom."

Suppose we choose $n = 180$. We then have a one-to-one correspondence between points on the semi-circle and the real numbers from 0 to 180. This is a coordinate system for the semi-circle.

We can now use this coordinate system to measure any angle, say $\angle pqr$. To do this we place our scale so that the center of the semi-circle c falls on the vertex of the angle, q, and the ray $\overrightarrow{qr}$ falls on the radius $\overline{cb}$.

The ray $\overrightarrow{qp}$ will then intersect the arc at some point with coordinate t, and we say that the measure of $\angle pqr$ is t.

If we decide to divide the semi-circle into 180 equal parts, then the unit of measurement is called the *degree*. We would thus say that $\angle pqr$ is a t degree (written $t°$) angle, or the measure of $\angle pqr$ in degrees is t. Since we will be using the degree exclusively as the unit of measurement, we will usually omit the word "degree" when giving the measure of an angle and write m $\angle pqr = t$. (Read, "the measure of $\angle pqr$ is t.")

Clearly, the choice of 180 is arbitrary. We could just as easily, for example, divide the semi-circle into 100 or 500 equal parts and make up some name for the units thus determined. This division of the half-circle into 180 parts comes to us from the ancient Babylonians. The Babylonian year was 360 days long—12 months of 30 days each. The apparently circular path of the sun was thus divided into 360 parts and from this we get the division of the semi-circle into 180 parts. The Babylonians also used a sexagesimal system (base 60) for recording large numbers. This probably accounts for the division of the degree into 60 minutes and each minute into 60 seconds.

To measure fractions of a degree, these subunits called minutes and seconds are used. Thus, for example, if the measure of an angle is a number halfway between 45 and 46, we would say that its measure was 45 degrees, 30 minutes, (written $45°30'$) rather than $45\frac{1}{2}$ degrees.

Radian Measure

Another way of measuring angles is by using a unit called the *radian*. To define a radian, consider a circle with center at c whose radius is one unit in length. (This is called the *unit circle*.) Now suppose we have a piece of string whose length is also one unit. Starting at some point a on the circle, place the string so that it follows the curve of the circle. The end of the string will be at some point b on the circle. The portion of the circle from a to b is called an *arc* and its length is equal to the length of the string —one unit.

In Figure 3-9 look at the angle bca which is the union of the two rays $\overrightarrow{ca}$ and $\overrightarrow{cb}$. This angle is determined by the arc whose length measures one unit, and $\angle bca$ is said to have measure *one radian*. In measuring an angle by radians, we are measuring a length of arc on the unit circle. Suppose, for example, the arc ad on the unit circle has length $1\frac{1}{2}$ units. Then, $\angle dca$ has measure $1\frac{1}{2}$ radians.

Since the circumference of a circle is 2π times its radius, the circumference of a unit circle is $2\pi \cdot 1 = 2\pi$ units, and the length of the semi-circular arc is half this, or π units. The length of arc associated with a 90° angle is thus $\pi/2$, and $\pi/2$ radians is equivalent to 90 degrees. One radian is equivalent to $180/\pi$ or approximately 57 degrees.

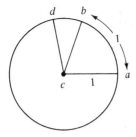

The measure of $\angle bca$ is 1 radian $\doteq 57°$.

Figure 3-9

Properties of Measure of Angles

The measure of angles has properties similar to the properties of the measure of line segments.

1. The measure of an angle in degrees is a positive real number between 0 and 180. This is a consequence of the way in which we have defined an angle as the union of two rays. We have no need of negative numbers or numbers larger than 180 to measure angles. In fact, since an angle is the union of two *distinct, non-opposite* rays, we don't need the numbers 0 or 180 to measure angles either. Our definition does not admit a straight line or a ray as an angle.

2. If p is a point in the interior of $\angle abc$, then

$$\text{m} \angle abp < \text{m} \angle abc$$

3. If p is a point in the interior of $\angle abc$, then

$$\text{m} \angle abp + \text{m} \angle pbc = \text{m} \angle abc$$

This is called the additive property of measure for angles.

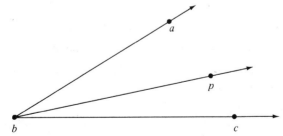

If p is a point in the interior of $\angle abc$, then
$$\text{m} \angle abp < \text{m} \angle abc$$
and $\quad\quad \text{m} \angle abp + \text{m} \angle pbc = \text{m} \angle abc$

Figure 3-10

We can use property 3 to find the measure of an angle which is not in the standard position on the degree scale. Consider the angle *dba* in Figure 3-11.

By property 3

$$\text{m} \angle dba + \text{m} \angle abc = \text{m} \angle dbc$$

Therefore

$$\text{m} \angle dba = \text{m} \angle dbc - \text{m} \angle abc$$
$$= 120 - 45$$
$$= 75$$

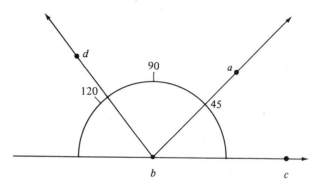

$$\text{m} \angle dba = \text{m} \angle dbc - \text{m} \angle abc = 75$$

Figure 3-11

Classification of Angles

We can now classify angles according to their measure.

An angle is called a *right angle* if its measure is 90.

An angle is *acute* if its measure is less than 90.

An angle is *obtuse* if its measure is greater than 90.

Every angle can thus be classified as right, acute, or obtuse, and these three classes are non-overlapping.

Two angles are *complementary* angles if the sum of their measures is 90. Each angle is said to be the complement of the other.

Two angles are *supplementary* angles if the sum of their measures is 180. In this case each is said to be the supplement of the other.

Clearly, if two angles are complementary, then they both must be acute angles. On the other hand, if two angles are supplementary, then except for one case, one of the angles must be acute and the other obtuse. What is the one exception?

Congruence of Angles

Now we can define a congruence relation between angles. Two angles will be said to be *congruent* if they have the same measure. Thus, all right angles are congruent.

At this point we must carefully distinguish between the "equals" relation and the "congruence" relation between angles. Since angles have been defined to be sets of points, equality between two angles will mean equality of the two point sets. Thus $\angle cab = \angle def$ if every point of $\angle def$ is a point of $\angle cab$ and every point of $\angle cab$ is a point of $\angle def$. In other words, $\angle cab = \angle def$ means that these are two different names for the same set.

Clearly, two angles can have the same measure (i.e., be congruent) but be two entirely different sets of points. To denote *congruence* of two angles we will use the symbol $\cong$. Thus, if $\angle cab$ is congruent to $\angle def$, we write $\angle cab \cong \angle def$, or we could say m $\angle cab =$ m $\angle def$. If two angles are equal, then they are congruent, but the converse of this statement is not true; two angles can be congruent, but not equal. Thus, in Figure 3-12 we would say $\angle cab = \angle dae$ and $\angle cab \cong \angle rst$; however, $\angle cab \neq \angle rst$.

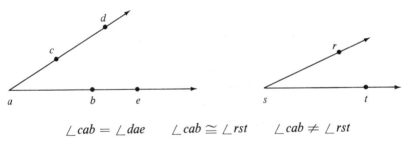

$$\angle cab = \angle dae \qquad \angle cab \cong \angle rst \qquad \angle cab \neq \angle rst$$

Figure 3-12

We will be very careful to use the word "equals" only where it applies. Thus, it would *not* be correct to say "all right angles are *equal*." It is true, however, that all right angles are *congruent*, since they all have the same measure, 90.

Exercise 3.4

1. Suppose that we subdivide the semi-circle into 100 equal parts and call the unit thus determined a *dex*. What is the measure in dexes

of an angle whose measure in degrees is 90? 45? 30? 135? Find a formula that will relate measurement in dexes to measurement in degrees.

$$1 \text{ degree} = \underline{\hspace{1cm}} \text{ dexes}$$
$$1 \text{ dex } = \underline{\hspace{1cm}} \text{ degrees}$$

2. What is the measure in radians of an angle whose measure in degrees is 45? 30? 60? 135? One degree is equivalent to how many radians?

3. What is the measure in degrees of an angle whose measure in radians is $\pi/2$? $\pi/4$? $\pi/6$? $\pi/3$? $2\frac{1}{2}$? 3?

4. Use the following figure to tell whether each of the statements are true or false.

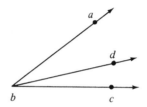

(a) m $\angle abd <$ m $\angle abc$.
(b) $\angle abd$ is a subset of $\angle abc$.
(c) m $\angle abd +$ m $\angle dbc =$ m $\angle abc$.
(d) $\angle abd \cup \angle dbc = \angle abc$.
(e) The interior of $\angle dbc$ is a subset of the interior of $\angle abc$.
(f) Interior of $\angle abc =$ interior of $\angle dbc \cup$ interior of $\angle abd$.

5. From the figure below find each of the following:

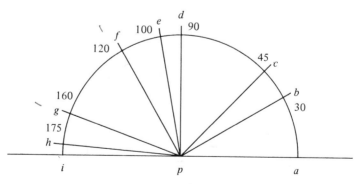

(a) m $\angle apc$ (b) m $\angle bpc$
(c) m $\angle dpb$ (d) m $\angle dph$

(e)   m $\angle epd$ (f)   m $\angle bpa$ + m $\angle dpe$

(g)   m $\angle cpg$ − m $\angle fpg$ (h)   m $\angle dph$ + m $\angle hpi$

(i)   m $\angle cpd$ + m $\angle apc$ + m $\angle dpi$

(j)   m $\angle epa$ − m $\angle hpg$

6.

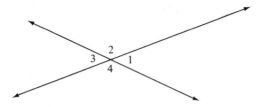

Given m $\angle 1 = 45$, find the measures of $\angle 2$, $\angle 3$, and $\angle 4$. Name two pairs of vertical angles in the figure. What conclusion do you reach about the measures of vertical angles?

7.   If two angles with the same measure are supplementary, what is their measure? If two angles with the same measure are complementary, what is their measure?

8.   What is the measure of an angle whose measure is twice the measure of its complement? What is the measure of an angle whose measure is twice the measure of its supplement?

9.   Show that the congruence relation between angles is an equivalence relation.

10.   Let S be the set of acute angles and let the relation Ⓡ be "is the complement of." Is this relation reflexive? Is it symmetric? Is it transitive? Explain.

5.   Perpendicularity

Two intersecting lines are said to be *perpendicular* if their union contains a right angle.

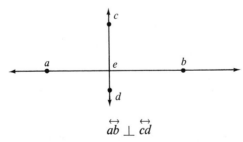

$$\overleftrightarrow{ab} \perp \overleftrightarrow{cd}$$

Figure 3-13

In Figure 3-13 we would say that since m $\angle ceb = 90$, then $\overleftrightarrow{ab}$ is perpendicular to $\overleftrightarrow{cd}$ and we write this

$$\overleftrightarrow{ab} \perp \overleftrightarrow{cd}$$

We can also define the relation "is perpendicular to" between two rays or between two line segments. Two rays will be said to be perpendicular if the lines containing them are perpendicular; and in the same way, two line segments are perpendicular if the lines containing them are perpendicular. Thus, two rays or line segments might be perpendicular even though their intersection were empty.

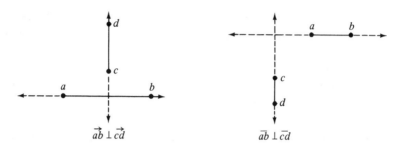

Figure 3-14

If we recall the definition of adjacent angles, we can describe perpendicular lines in still another way. Two angles are adjacent if they have the same vertex and their intersection is a ray. We might say that two intersecting lines are perpendicular if their union contains congruent adjacent angles.

6. Vertical Angles

Now that we have defined congruence of angles, we are ready to prove some simple theorems about congruences of certain types of angles. Recall that two angles are supplementary if the sum of their measures is 180, and each is called the supplement of the other. The following theorem is easy.

THEOREM 1.   If two angles are congruent, then their supplements are congruent.

   Proof: Suppose $\angle a \cong \angle b$. Then, m $\angle a =$ m $\angle b$ and this measure is some number, call it n, between 0 and 180.

Then the supplement of $\angle a$ must have measure $180 - n$. Similarly, the supplement of $\angle b$ must have measure $180 - n$. Since these supplements have the same measure they are congruent. ∎

Since every angle is congruent to itself (Why?), it follows that if two angles are supplements of the same angle, then they must be congruent.

We recall that two angles are called *vertical* angles if their union is a pair of intersecting lines.

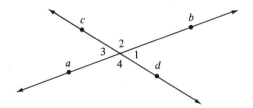

Vertical angles are congruent.

Figure 3-15

Thus, in Figure 3-15 $\angle 1$ and $\angle 3$ are vertical angles, as are $\angle 2$ and $\angle 4$. The following theorem is very useful. To prove it we will use Theorem 1.

THEOREM 2.   Vertical angles are congruent.

Proof: Consider the vertical angles, $\angle 1$ and $\angle 3$. Since $\overleftrightarrow{cd}$ is a line, $\angle 2$ is supplementary to $\angle 1$, and since $\overleftrightarrow{ab}$ is a line, $\angle 2$ is supplementary to $\angle 3$. Thus, both $\angle 1$ and $\angle 3$ have the same supplementary angle, $\angle 2$, and since $\angle 2$ is certainly congruent to itself, it follows from Theorem 1 that $\angle 1$ is congruent to $\angle 3$. ∎

7. Classification of Triangles

We can classify triangles according to the measures of their sides.

A triangle is *isosceles* if at least two of its sides are congruent.

A triangle is *equilateral* if all three sides are congruent.

A triangle is *scalene* if no two sides are congruent.

According to these definitions an equilateral triangle is isosceles, but a triangle may be isosceles and not be equilateral. (Figure 3-16) A scalene triangle is neither isosceles nor equilateral.

We can picture these relationships in Figure 3-17. The large circle here denotes the set of all triangles.

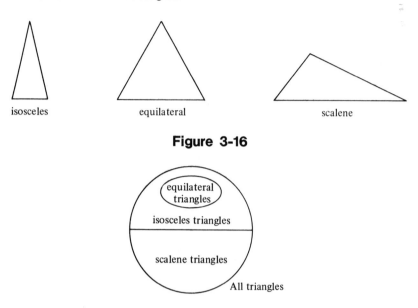

isosceles   equilateral   scalene

Figure 3-16

Figure 3-17

We can also classify triangles by the measures of one or more of their angles.

A triangle is called a *right* triangle if one of its angles is a right angle.

A triangle is *obtuse* if one of its angles is an obtuse angle.

A triangle is *acute* if all three of its angles are acute.

A triangle is *equiangular* if all three of its angles are congruent.

We can picture this situation in the diagram in Figure 3-18.

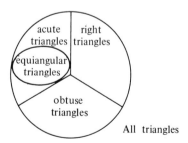

Figure 3-18

Exercise 3.7

1.

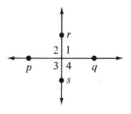

Given $\overleftrightarrow{pq} \perp \overleftrightarrow{rs}$. According to the definition, $\overleftrightarrow{pq} \cup \overleftrightarrow{rs}$ contains a right angle. Suppose that this right angle is $\angle 1$.

(a) What can you say about the measure of $\angle 2$? Why?

(b) What can you say about the measure of $\angle 3$? Why?

(c) What can you say about the measure of $\angle 4$? Why?

(d) How many right angles are formed by two perpendicular lines?

2.

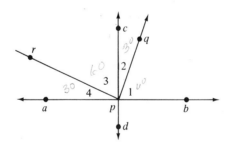

Given $\overleftrightarrow{ab} \perp \overleftrightarrow{cd}$, $\overrightarrow{pq} \perp \overrightarrow{pr}$, m $\angle 4 = 30$. Find m $\angle 2$, m $\angle 3$, m $\angle 1$.

3. In the figure of problem 2, name as many pairs of complementary angles as you can. Name as many pairs of adjacent angles, congruent angles, and supplementary angles as you can.

4.

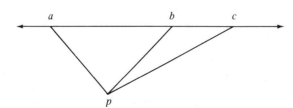

Copy the figure above carefully. Now find by eye a point x on $\overleftrightarrow{ac}$ such that $\overleftrightarrow{px} \perp \overleftrightarrow{ac}$. Measure carefully the line segments $\overline{pa}$, $\overline{pb}$, $\overline{pc}$, and $\overline{px}$. Which is the smallest? If you had to measure the distance from the point p to the line $\overleftrightarrow{ac}$, which of the above measurements would you choose? Make up a definition for the distance from a point p to a line L.

5. Prove that if two acute angles are congruent, then their complements are congruent.

6. Prove that if $\angle b$ and $\angle c$ are both supplementary to $\angle a$ then $\angle b \cong \angle c$.

7.

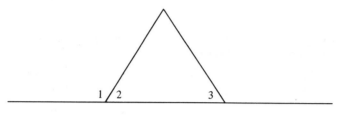

In the figure above if $\angle 1$ and $\angle 3$ are supplementary, what can you say about $\angle 2$ and $\angle 3$? Why?

8.

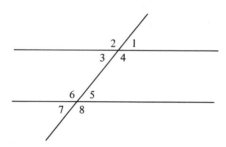

In the figure above, if m $\angle 1 = 30$ and $\angle 1 \cong \angle 7$, find the measures of angles 2, 3, 4, 5, 6, 7, and 8. Give reasons.

9.

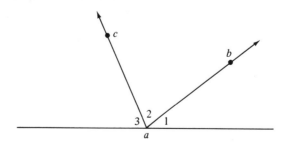

In the figure above $\angle 1$ and $\angle 3$ are complementary. What is the measure of $\angle 2$? Why? What can you say about the relationship between $\overrightarrow{ab}$ and $\overrightarrow{ac}$?

10. Let S be the set of all lines, and let ⓇＲ be the relation "is perpendicular to." Is this an equivalence relation? Why or why not?

11. If two vertical angles are also supplementary, what is their measure? Why? Sketch.

12. Can an isosceles triangle be a right triangle? Can it be acute? Can it be obtuse? Illustrate your answer.

13. Can an equilateral triangle be a right triangle? Acute? Obtuse? Illustrate your answer.

14. Can a scalene triangle be a right triangle? Acute? Obtuse? Illustrate your answer.

New Terms Found in This Chapter

Term	Section	Term	Section
number line	2	complementary angles	4
coordinate system	2	supplementary angles	4
coordinate	2	congruence of angles	4
midpoint	2	perpendicular	5
congruence of line segments	2	vertical angles	6
circle	3	isosceles triangle	7
radius	3	equilateral triangle	7
diameter	3	scalene triangle	7
degree	4	right triangle	7
radian	4	obtuse triangle	7
right angle	4	acute triangle	7
acute angle	4	equiangular triangle	7
obtuse angle	4		

4: Congruence of Plane Figures

1. Congruence of Polygons

In the last chapter we defined a congruence relation between line segments and between angles. Now we will extend this idea of congruence to other plane figures, particularly triangles. Since a triangle is the union of line segments, and since every triangle has three angles closely associated with it, it seems reasonable to use the notions of congruence of line segments and of angles to define congruence of triangles.

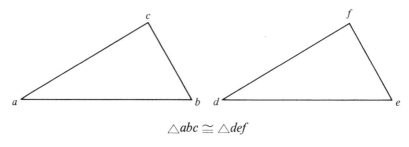

$$\triangle abc \cong \triangle def$$

Figure 4-1

In Figure 4-1, if $\triangle abc$ were traced on thin paper and the paper then placed over $\triangle def$ so that vertex a fell on vertex d, c on f, and b on e, the two figures would coincide point for point, and we would say that $\triangle abc$ was *congruent* to $\triangle def$. However, if we tried to place $\triangle abc$ on $\triangle def$ with vertex a lying on vertex e, b on f, and c on d, the two figures would not coincide.

This suggests that if we want to describe a congruence relation between two triangles, we must specify a particular one-to-one correspondence between the vertices of the two triangles. If we have a correspondence between the vertices of two triangles, clearly this determines a correspondence between their sides and angles. The correspondence $a \leftrightarrow d$, $b \leftrightarrow e$, $c \leftrightarrow f$ determines the correspondence $\overline{ab} \leftrightarrow \overline{de}$, $\overline{ac} \leftrightarrow \overline{df}$, $\overline{cb} \leftrightarrow \overline{fe}$

between the sides of the triangles; and the correspondence $\angle a \leftrightarrow \angle d$, $\angle b \leftrightarrow \angle e$, $\angle c \leftrightarrow \angle f$ between their angles.

We can describe the one-to-one correspondence by naming the vertices of each triangle in the order desired. Thus, if we want to describe the one-to-one correspondence between $\triangle abc$ and $\triangle def$ which will match vertex a to vertex d, vertex b to vertex e, and vertex c to vertex f, we can write this

$$abc \longleftrightarrow def$$

The order in which the vertices are named gives the correspondence. Thus, we have

We will say then that two triangles are *congruent* if we can find a one-to-one correspondence between their vertices so that corresponding sides are congruent and corresponding angles are congruent. This one-to-one correspondence will be called a *congruence*.

In Figure 4-1 the two triangles are congruent under the correspondence $abc \leftrightarrow def$ since

$$\overline{ab} \cong \overline{de} \qquad\qquad \angle a \cong \angle d$$
$$\overline{ac} \cong \overline{df} \quad \text{and} \quad \angle b \cong \angle e$$
$$\overline{cb} \cong \overline{fe} \qquad\qquad \angle c \cong \angle f$$

and we write this $\triangle abc \cong \triangle def$.

Observe that our notation indicates what the congruence is. Thus, when we write $\triangle abc \cong \triangle def$, we mean that $abc \leftrightarrow def$ is the correspondence between vertices which matches congruent sides and congruent angles.

In Figure 4-2, $\overline{ab} \cong \overline{de}$, $\overline{ac} \cong \overline{df}$, $\overline{bc} \cong \overline{ef}$, $\angle a \cong \angle d$, $\angle b \cong \angle e$, and $\angle c \cong \angle f$. Corresponding congruent parts are indicated by the same number of marks. A right angle will be indicated as in Figure 4-3.

In this definition, we refer to congruence of "two triangles." However, this does not mean that we must have two *different* triangles in order to talk about congruence. Every triangle is, in fact, congruent to itself under the correspondence that matches each vertex to itself.

We can easily extend this definition of congruence of triangles to congruence of polygons. Thus, we can say that two polygons $p_1 p_2 p_3 \ldots p_n$ and $q_1 q_2 q_3 \ldots q_n$ are congruent if we can find a one-to-one correspondence between their vertices so that corresponding sides are congruent and

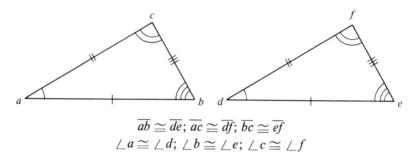

$$\overline{ab} \cong \overline{de};\ \overline{ac} \cong \overline{df};\ \overline{bc} \cong \overline{ef}$$
$$\angle a \cong \angle d;\ \angle b \cong \angle e;\ \angle c \cong \angle f$$

Figure 4-2

A right angle

Figure 4-3

corresponding angles are congruent. In Figure 4-4 the two polygons are congruent under the correspondence $p_1p_2p_3p_4p_5p_6 \longleftrightarrow q_1q_2q_3q_4q_5q_6$.

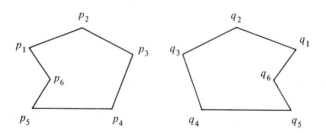

Polygon $p_1p_2p_3p_4p_5p_6$ is congruent to
polygon $q_1q_2q_3q_4q_5q_6$.

Figure 4-4

2. Congruence Theorems for Triangles

If we have a correspondence between two triangles that we think is a congruence, we do not have to verify that all six corresponding parts are congruent. In what follows we will investigate just how much information we must have about our correspondence in order to conclude that it is a congruence.

Suppose that we have a triangle *abc*, and we wish to copy it (construct another triangle which will be congruent to △*abc*). To do this we will use the straight-edge, the compass and the protractor.*

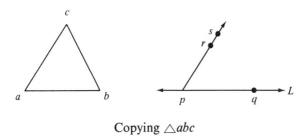

Copying △*abc*

Figure 4-5

First with the straight-edge we draw a line *L* on which we choose a point *p*, which we wish to correspond to point *a*. Next, we set the compass to the length *ab*, place the point of the compass at *p* and mark off the length *pq* on the line *L*. Point *q* will correspond to the point *b*. From our construction we have $\overline{ab} \cong \overline{pq}$.

Next, we use the protractor to construct at *p* an angle having the same measure as ∠*a*. Thus, ∠*p* ≅ ∠*a*.

Now using the compass again, we mark off on the ray $\overrightarrow{ps}$ a segment $\overline{pr}$ so that $\overline{ac} \cong \overline{pr}$. The point *r* will, of course, correspond to the point *c*.

Now at this point, although we have only copied two sides and the angle between them, it is clear that triangle *pqr* is completely determined. That is, all that remains is to draw the segment $\overline{rq}$ with the straight-edge, and there is only one way in which this can be done. Clearly, △*abc* ≅ △*pqr*.

This illustrates our basic congruence axiom.

SAS Axiom

Two triangles are congruent if we can find a one-to-one correspondence between their vertices so that two sides and the included angle of the first triangle are congruent to the corresponding sides and included angle of the second.

Here SAS stands for "Side, Angle, Side."

Thus, in Figure 4-6, if we know that ∠*a* ≅ ∠*f*, $\overline{ab} \cong \overline{fe}$, and $\overline{ac} \cong \overline{fd}$, then by the SAS Axiom, the correspondence *abc* ↔ *fed* is a congruence; △*abc* ≅ △*fed*.

---

*In Chapter 5 these constructions will be done with straight-edge and compass alone.

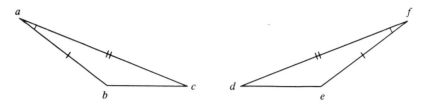

$\triangle abc \cong \triangle fed$ by the SAS Axiom.

Figure 4-6

The SAS Axiom states that two sides and an included angle uniquely determine a triangle. The following axiom says that two angles and an included side are sufficient to determine a triangle.

Suppose that we set about copying $\triangle abc$ in another way.

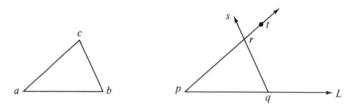

Copying $\triangle abc$ in a second way

Figure 4-7

With a straight-edge we draw a line L as before and mark off the line segment $\overline{pq}$ congruent to $\overline{ab}$. Now using the protractor, we construct at p an angle congruent to $\angle a$. Next, on the same side of L we construct at q an angle congruent to $\angle b$. The two rays $\overrightarrow{pt}$ and $\overrightarrow{qs}$ will intersect in a point r, and this determines uniquely the triangle pqr. This illustrates our second congruence axiom.

ASA Axiom

Two triangles are congruent if we can find a one-to-one correspondence between their vertices so that two angles and the included side of the first are congruent to the corresponding angles and included side of the second.

Here ASA stands for "Angle, Side, Angle."

In the triangles in Figure 4-8, since $\angle a \cong \angle d$, $\angle b \cong \angle e$, $\overline{ab} \cong \overline{de}$, we can conclude that $\triangle abc \cong \triangle def$.

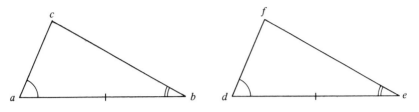

$\triangle abc \cong \triangle def$ by the ASA Axiom.

Figure 4-8

We will now copy $\triangle abc$ in yet a third way.

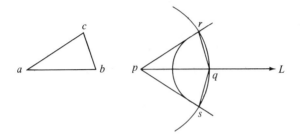

Copying $\triangle abc$ in a third way

Figure 4-9

As before we mark off line segment $\overline{pq}$ congruent to $\overline{ab}$. Now we set the compass so that one point is at a and the other at c. With the compass at this setting we draw a circle whose center is at p. The radius of this circle has measure ac. Now in the same way we draw a circle having center at q and radius bc. These circles will intersect at two points, r and s, one above and one below the line L. Clearly, $\overline{pr} \cong \overline{ac}$ and $\overline{ps} \cong \overline{ac}$; moreover, $\overline{qr} \cong \overline{bc}$ and $\overline{qs} \cong \overline{bc}$.

The three sides of the triangles pqr and pqs are congruent, respectively, to the three sides of $\triangle abc$. However, we have obtained not one triangle but two! Does this mean that given the measures of three sides of a triangle we can construct two different triangles congruent to it? Not really. Consider what would happen if the figure were folded along line L. The point s would fall on the point r, and we can see that the two triangles will be congruent under the correspondence $pqr \leftrightarrow pqs$. From this construction we conclude that a triangle is uniquely determined by the lengths of its sides, and this illustrates our third congruence axiom.

SSS Axiom

Two triangles are congruent if we can find a one-to-one correspondence between their vertices so that all three pairs of corresponding sides are congruent.

SSS stands for "Side, Side, Side."

A carpenter is familiar with the fact that three sides determine one and only one triangle. If he wants a certain angle to be rigid, he nails a brace across the two sides of the angle, making the angle a part of the unique triangle formed by the three fixed sides.

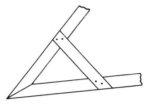

A structural application of the SSS Axiom

Figure 4-10

These three congruence axioms are very old. They appear in Book I of Euclid's *Elements* and probably were known before Euclid. The SAS Axiom occurs very early. It is Proposition #4, and Euclid does not state it as an Axiom but gives a "proof" of it. The proof which Euclid gives is similar to the discussion given at the beginning of this section on copying a triangle given two sides and the included angle. He "picks up" one triangle and superimposes it on the second and observes that if this is done so that the equal sides and equal angles coincide, then the other side and angle must coincide also.

In proving this theorem Euclid does not refer to any of the Axioms he has listed earlier; however, it is clear that he is assuming a fact that he has not listed as an axiom—that a geometric figure can be moved without changing its shape or size.

This proof of Euclid's is rather unsatisfactory, depending as it does on an unstated axiom about movability. It also depends rather heavily on a picture, and mathematicians have learned to be wary of so called "proofs" that are based on a figure.* The result is that today Euclid's "proof" is not considered to be a proof at all, and in more modern geometry textbooks the SAS statement is given as an Axiom, accepted without proof.

---

*See for example the "proof" that every triangle is isosceles in E. A. Maxwell, *Fallacies in Mathematics* (New York: Cambridge University Press, 1963).

That is not to say that the SAS statement *cannot* be proved. If we make the proper assumptions—take certain statements as axioms, then use those axioms—it certainly could be proved. However, it cannot be proved using only the axioms which Euclid listed. The question now arises, if we must make additional assumptions in order to prove the SAS statement, why not just assume the SAS statement itself? This is what we have done. The ASA and SSS Axioms then become theorems and can be proved using the SAS statement as an axiom. For details of these proofs, the student is referred to E. Moise, *Elementary Geometry from an Advanced Standpoint* (Reading, Mass.: Addison-Wesley Publishing Co., Inc., 1963), pp. 86–88.

Note that in each of these statements we had a congruence between at least one pair of corresponding sides. Congruence of corresponding angles is not enough to guarantee congruence of the triangles. In other words, we have no "AAA Axiom" or "AAA Theorem." The student should satisfy himself that this is the case by drawing two triangles which are not congruent, but for which all three pairs of corresponding angles are congruent. Such triangles have the same shape, but are not necessarily the same size. They are called "similar" triangles and will be studied in a later chapter.

At this point we raise the question whether or not it is necessary in the SAS Axiom for the given angle to be *included* between the two sides, and whether or not it is necessary in the ASA Axiom for the given side to be *between* the two angles. In other words, is there an SSA Axiom or an AAS Axiom?

In a later chapter will see that the answer to the second question is "yes." That is, if two angles and a side of one triangle are congruent to two angles and a side of a second, then the two triangles are congruent whether the side is between the two angles or not.

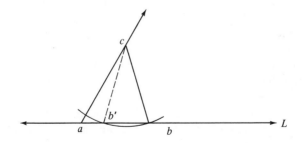

$\triangle abc$ and $\triangle ab'c$ have two sides and a non-included angle congruent, but they are not congruent.

Figure 4-11

However, it is essential that the angle in the SAS Axiom be the *included* angle. The following construction will illustrate this. (Figure 4-11)

First, we construct an angle, $\angle a$. Next we measure off length ac. Now with the point of the compass at c, we can draw an arc cutting line L in two points, b and b'. Thus, $\overline{cb} \cong \overline{cb'}$. Clearly $\triangle acb'$ is not congruent to $\triangle acb$; however two sides and an angle of $\triangle acb'$ are congruent to two sides and an angle of $\triangle acb$.

Earlier we defined an isosceles triangle as one having two congruent sides. We can now use the SAS Axiom to prove an interesting property of isosceles triangles: If two sides of a triangle are congruent, then the angles opposite these sides are congruent. Sometimes you will see this stated "the base angles of an isosceles triangle are congruent." The "base angles" of an isosceles triangle are simply the angles opposite the congruent sides. The adjective "base" does *not* refer to a location at the bottom of the figure. Thus in each of the triangles in Figure 4-12 $\angle a$ and $\angle c$ are base angles.

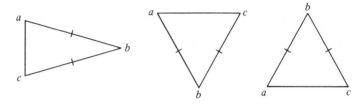

$\angle a$ and $\angle c$ are base angles of these isosceles triangles.

Figure 4-12

The following theorem was considered very difficult in medieval universities. It was called the *pons asinorum* or "bridge of asses," probably because the less able students could not understand it and could proceed no further in the study of geometry. Euclid's proof of this theorem was considerably longer and more complicated than the one which follows. To prove this theorem we will show that $\triangle abc$ is congruent to itself under the particular correspondence $cab \longleftrightarrow cba$.

To picture this correspondence, think of tracing $\triangle abc$ on thin paper. Now pick the tracing paper up, turn it over and replace it on the original triangle so that vertex b falls on vertex a, a on b, and c on c. This correspondence will be a congruence.

THEOREM 1. If two sides of a triangle are congruent, then the angles opposite these sides are congruent.

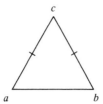

$\triangle abc$ is congruent to itself under the correspondence $abc \longleftrightarrow bac$.

Figure 4-13

Proof:

Let $\triangle abc$ be an isosceles triangle with $\overline{ac} \cong \overline{bc}$. We want to show that $\angle a \cong \angle b$. To do this, consider the correspondence $cab \longleftrightarrow cba$.

Under this correspondence $\angle c \longleftrightarrow \angle c$ and $\overline{ac} \longleftrightarrow \overline{bc}$. Since $\overline{ac} \cong \overline{bc}$, $\overline{bc} \cong \overline{ac}$, and $\angle c \cong \angle c$, under this correspondence, two sides and the angle between them are congruent to their corresponding parts. Thus, by the SAS Axiom $\triangle cab \cong \triangle cba$. Since $\angle a$ corresponds to $\angle b$ under the congruence, we must have $\angle a \cong \angle b$. ∎

It follows from this that in an equilateral triangle, all three angles are congruent; i.e., an equilateral triangle is equiangular.

Theorem 2 is the converse of Theorem 1 and can be proved in much the same way, using the ASA Axiom.

THEOREM 2   If two angles of a triangle are congruent, then the sides opposite these angles are congruent.

From this theorem it follows that in an equiangular triangle, all three sides are congruent (an equiangular triangle is equilateral).

3. Congruence of Figures in the Plane

Although we have defined congruence of triangles in terms of congruence of line segments and angles, it is clear that we can have congruences between plane figures which are not the union of line segments. For example, two circles are congruent if their radii are congruent.

In defining a congruence between two triangles, we specified a one-to-one correspondence between vertices. It is not hard to see that such a correspondence does, in fact, define a one-to-one correspondence between all of the points of the triangles.

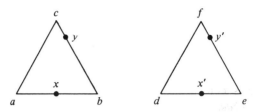

The one-to-one correspondence $abc \leftrightarrow def$ between vertices determines a correspondence between all the points of the triangles.

Figure 4-14

Thus, in Figure 4-14 if $\triangle abc \cong \triangle def$, then the midpoint x of $\overline{ab}$ will correspond to the midpoint x' of $\overline{de}$; and in general, to any point y on $\triangle abc$ there will correspond one and only one point, y', on $\triangle def$.

Congruences between triangles have a very important property. They *preserve distances*. We have already seen that if vertex a corresponds to vertex d and if b corresponds to e, then $\overline{ab} \cong \overline{de}$ (the distance from a to b is the same as the distance from d to e).

We can make an even stronger statement. If x and y are any two points on $\triangle abc$ and x' and y' are the points corresponding to them on $\triangle def$ under a congruence, then the distance from x to y is the same as the distance from x' to y'. This is what we mean when we say that congruence preserves distances.

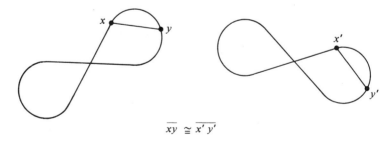

$$\overline{xy} \cong \overline{x'y'}$$

Two plane figures are congruent if there is a one-to-one correspondence between their points that preserves distances.

Figure 4-15

With this in mind we can now extend the notion of congruence to plane figures in general. If we have two plane curves, we will say that they are congruent if there is a one-to-one correspondence between their points that preserves distances.

Exercise 4.3

1. Let S be the set of all triangles, and Ⓡ be the relation "is congruent to." Show that Ⓡ is an equivalence relation.

2. Use the following figure to give an alternative proof of Theorem 1.

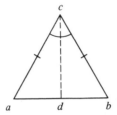

The line segment $\overline{cd}$ has been drawn so that it bisects angle c, so you may assume that $\angle acd \cong \angle dcb$. Note that this proof depends on the picture in that you must assume that d lies between a and b.

3. In problem 1 you noted that every triangle is congruent to itself. Consider the equilateral triangle abc.

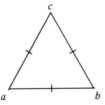

If we choose the correspondence $abc \longleftrightarrow abc$, clearly this defines a congruence. Find two more correspondences between the vertices of $\triangle abc$ that are congruences.

4. In each of the following figures there is at least one pair of congruent triangles. Identify as many congruences as you can, using the proper notations. You may assume in this problem only that appearances are not deceiving. If two segments or two angles look as though they are congruent, they are.

(1)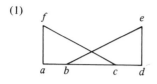

(2)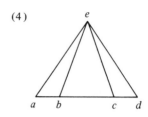

(3)

(4)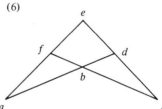

(5)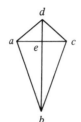

(6)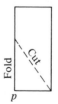

5. The SAS Axiom tells us that two sides and the angle between them uniquely determine a triangle. Is this true for a quadrilateral? Draw diagrams to illustrate your answer. How many angles and/or sides do you think it would take to uniquely determine a quadrilateral? State a theorem for quadrilaterals that is similar to the SAS Axiom; to the ASA Axiom.

6. Fold a sheet of notebook paper in half. Cut off a section as shown and unfold the piece you have cut off. What kind of triangle is it? Label

the vertices and the point *p*. Use this paper to demonstrate the fact that every isosceles triangle has associated with it two congruent right triangles.

7. The SSS Axiom tells us that the sides of a triangle uniquely determine that triangle. Is a corresponding statement true for quadri-

laterals? That is, is a quadrilateral uniquely determined by its four sides? Explain and illustrate.

8. (a) Given any three positive real numbers, *a*, *b*, and *c*, can you construct a triangle having sides whose measures are *a*, *b*, and *c*?

    (b) Can you construct a triangle whose sides measure 2, 3, and 5 units, respectively? 2, 3, and 6?

    (c) What restrictions, if any, must you place on *a*, *b*, and *c* in order that they might represent the measures of the sides of a triangle?

9. Given any four positive real numbers can you construct a quadrilateral having sides whose measures are these four numbers? What restrictions, if any, must you place on four numbers in order that they might represent the measures of the sides of a quadrilateral?

10. Given any three positive real numbers, *a*, *b*, and *c*, with *b* and *c* less than 180, can you construct a triangle having two angles whose measures are *b* and *c*, respectively, and the included side having measure *a*? Illustrate.

11. Given any three positive real numbers, *a*, *b*, and *c*, with *c* less than 180, can you construct a triangle having two sides whose measures are *a* and *b* with the included angle having measure *c*? Illustrate.

12. If you know that two triangles are equilateral, does it follow that they are congruent? If your answer is no, what additional fact would you need to show that they are congruent?

13. Let us define the *perimeter* of a triangle to be the sum of the measures of its sides. If two triangles are congruent, do they have the same perimeter? If two triangles have the same perimeter are they necessarily congruent? Illustrate your answer.

14. Let *abc* be an equilateral triangle, $\overline{ab} \cong \overline{bc} \cong \overline{ca}$. Show that $\angle a \cong \angle b \cong \angle c$.

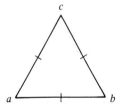

15. Prove Theorem 2.

16. Let $\triangle abc$ be an equiangular triangle, $\angle a \cong \angle b \cong \angle c$. Show that $\overline{ab} \cong \overline{ac} \cong \overline{bc}$.

17. You have proved that every equilateral triangle is equiangular and

every equiangular triangle is equilateral. What do you conclude about the set of all equilateral triangles and the set of all equiangular triangles?

18. Suppose that you have a quadrilateral *abcd* whose opposite sides are congruent, i.e., $\overline{ab} \cong \overline{cd}$ and $\overline{bc} \cong \overline{ad}$. Prove that $\angle a \cong \angle c$. (*Hint :* Draw a diagonal.)

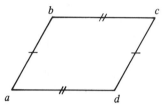

19. If $\triangle abc$ is isosceles with $\overline{ab} \cong \overline{bc}$, prove that $\angle 1 \cong \angle 2$.

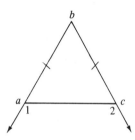

20. In the following figures decide if the given information is sufficient to prove that the shaded triangles are congruent. Do not assume anything about these figures except the congruences which are indicated. Give reasons for your answers.

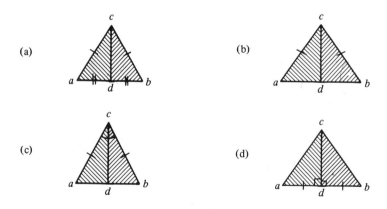

(a)

(b)

(c)

(d)

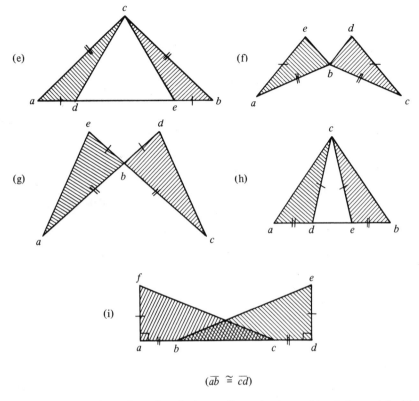

$(\overline{ab} \cong \overline{cd})$

21. Every equiangular triangle is equilateral. Does this statement hold for a quadrilateral? (If a quadrilateral has all four angles congruent, are all four sides congruent?) Sketch.

22. Draw a triangle with two angles not congruent. Compare the measures of the two sides opposite these angles. Which of the sides has the larger measure? Repeat for several other triangles. State your conclusion in the form of a theorem.

23. Draw a triangle with two sides not congruent. Compare the measures of the two angles opposite these sides. Which of the angles seems to have a larger measure? Repeat for several other triangles. State your conclusion in the form of a theorem.

24. Every equilateral triangle is equiangular. Is this true for quadrilaterals? (If a quadrilateral has all four sides congruent, are all four angles congruent?) This is equivalent to asking whether or not a quadrilateral made of four boards fastened together with a single nail through each vertex is rigid.

New Terms Found in This Chapter

Term	Section
congruent triangles	1
congruence	1
base angles of a triangle	2
congruence of plane figures	3
perimeter	Exercise 4.3

5: Geometric Constructions

1. Introduction

In earlier chapters you have drawn many sketches to aid your understanding. If you wanted to draw two congruent line segments, you probably used a ruler. To draw two congruent angles you used a protractor. The ruler and protractor are *measuring* instruments. We place them on geometric figures and read numbers from them.

In this chapter we will construct geometric figures using only an unmarked ruler or straight-edge and a compass. These are *not* measuring instruments. With the straight-edge we can draw a line segment, but we cannot measure its length. With the compass we can draw circles or portions of circles called arcs given a center and a radius.

Nevertheless, we can use these instruments to perform an amazing variety of constructions. We can, for instance, copy an angle or find the midpoint of a line segment without measuring.

Use of the straight-edge and compass in constructions goes back to the ancient Greeks. It is interesting to speculate why they used these two instruments and these two only. The Greeks considered the circle and the line to be perfect figures. Since the straight-edge is used to draw lines and the compass to draw circles, these instruments were acceptable. On the other hand, instruments for measuring were tools of the artisan and in the aristocratic Greek society were regarded as vulgar and unworthy of the attention of the mathematician-philosopher.

The Greek mathematician was interested in the logical problem of what kinds of figures could be drawn with these two instruments and not in any practical application of his figures. He valued knowledge for its own sake and regarded with contempt those who made practical use of their knowledge. The story is told of the pupil of Euclid who asked of what practical use his studies were. It is said that the master called a slave and commanded, "Give this youth a coin, so that he may derive gain from his knowledge."

Why geometric constructions? Are we just playing games with the straight-edge and compass? Hardly. Constructions are important because they allow us to *do* something which aids us in *thinking* about geometry. Using these instruments we can construct new figures which are combinations of circles and line segments and deduce theorems about them. In many theorems an important step in the proof involves the construction of an auxiliary line segment or arc.

A geometric construction is really a special sort of theorem. We start with certain assumptions or axioms and use these to construct a figure having certain properties. Once the figure is drawn we must justify our construction, that is, prove that the figure has the properties we claim.

Our construction axioms are related to the instruments we use in the construction. Since we are allowed to use the straight-edge, the first axiom tells us how we may use it.

CONSTRUCTION AXIOM 1   Given any two points a line may be drawn containing them.

The second construction axiom describes the use of the compass.

CONSTRUCTION AXIOM 2   Given any point and any line segment, a circle may be drawn with center at the point and radius congruent to the segment.

These axioms give us the rules that we must follow in drawing geometric figures. To prove that the figure has the desired properties, we can, of course, use any of the other axioms or theorems we have stated in previous chapters.

Note that the following constructions consist of various steps using straight-edge and compass, followed by a proof that the figure satisfies the requirements. In the following sections, the student should actually perform each construction as he studies it.

2. Elementary Constructions

Construction 1

Copying a line segment

Given line segment $\overline{pq}$, construct a second segment $\overline{p'q'}$ congruent to it.

First with the straight-edge draw a ray $\overrightarrow{p'r}$. Then, set the compass on $\overline{pq}$ so that one point of the compass is at p, and the second is at q. Now with one point of the compass at p', construct an arc with center at p' and radius pq. Call the intersection of the circle with the ray $\overrightarrow{p'r}$ the point q'. Then $\overline{pq} \cong \overline{p'q'}$.

Copying a line segment

Figure 5-1

Construction 2

Copying an angle
Given an angle, $\angle abc$, construct an angle congruent to it.

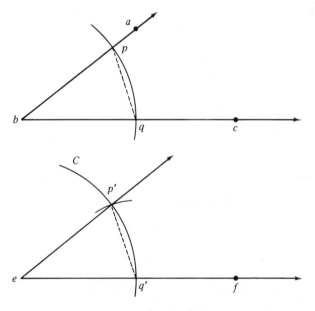

Copying an angle

Figure 5-2

1. With the straight-edge draw a ray $\overrightarrow{ef}$.
2. With the compass at any convenient setting construct a portion of a circle with center at b, cutting the rays $\overrightarrow{ba}$ and $\overrightarrow{bc}$ in the points p and q.
3. Draw an arc of a circle C having the same radius with center at e. This arc intersects the ray $\overrightarrow{ef}$ in the point q'. Then, $\overline{bq} \cong \overline{eq'}$.

4. Draw an arc with center at q' and radius equal to pq cutting C in point p'. Then, $\overline{pq} \cong \overline{p'q'}$.
5. With the straight-edge draw the ray $\overrightarrow{ep'}$.

Now we have constructed an angle at e. We must prove that this angle is congruent to $\angle abc$. To do this we will prove that $\triangle pbq$ is congruent to $\triangle p'eq'$.

Since p' and q' are two points on a circle with center at e, $\overline{ep'} \cong \overline{eq'}$. For the same reason, $\overline{bp} \cong \overline{bq}$. Now since $\overline{bq} \cong \overline{eq'}$, we conclude that $\overline{bp} \cong \overline{ep'}$. (What property of the congruence relation are we using here?) Moreover, from step 4, $\overline{pq} \cong \overline{p'q'}$.

Thus, $\triangle pbq \cong \triangle p'eq'$ by the SSS Axiom, and since $\angle b$ corresponds to $\angle e$ under this congruence, we conclude that $\angle b \cong \angle e$. ∎

Construction 3

Copying a triangle

If we wish to copy a triangle there are three methods we might use, depending on whether we wish to use the SAS Axiom, the ASA Axiom, or the SSS Axiom.

(i) Copy $\triangle abc$ using the SAS Axiom.

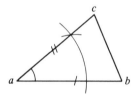 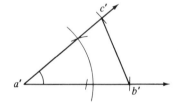

Copying a triangle using the SAS Axiom

Figure 5-3

1. Construct a segment $\overline{a'b'}$ congruent to $\overline{ab}$ by Construction 1.
2. Construct an angle at a' congruent to $\angle a$ by Construction 2.
3. Construct a segment $\overline{a'c'}$ congruent to $\overline{ac}$ by Construction 1.
4. Use the straight-edge to draw $c'b'$.

Then, $\triangle abc \cong \triangle a'b'c'$ by the SAS Axiom.

(ii) Copy $\triangle abc$ using the ASA Axiom.

This construction is left as a problem for the student. (See Chapter 4.)

(iii) Copy $\triangle abc$ using the SSS Axiom.

This construction is left as a problem for the student. (See Chapter 4.)

Construction 4

Bisecting an angle

An angle, $\angle abc$, is *bisected* by the ray $\overrightarrow{bd}$ if $\angle abd \cong \angle dbc$.

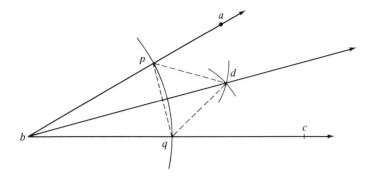

Bisecting an angle

Figure 5-4

Given an angle $\angle abc$, we wish to construct the bisector $\overrightarrow{bd}$.
1. With the compass set for any convenient radius, construct an arc with center at b cutting the rays $\overrightarrow{ba}$ and $\overrightarrow{bc}$ in the points p and q.
2. With centers at p and q construct two arcs having the same radius r, where r must be greater than $\frac{1}{2}pq$. (Why?) These two arcs will intersect in a point d in the interior of $\angle abc$.
3. With the straight-edge draw ray $\overrightarrow{bd}$.

Now we must show that $\overrightarrow{bd}$ really is the bisector of $\angle abc$ (that $\angle abd \cong \angle dbc$).

We note that

$$\overline{bp} \cong \overline{bq} \quad \text{(Why?)}$$
$$\overline{pd} \cong \overline{qd} \quad \text{(Why?)}$$
and
$$\overline{bd} \cong \overline{bd} \quad \text{(Why?)}$$

Then, $\qquad \triangle bpd \cong \triangle bqd \quad \text{(Why?)}$
and $\qquad \angle abd \cong \angle dbc \quad \text{(Why?)}$ ∎

Construction 5

Constructing the perpendicular bisector of a line segment
The perpendicular bisector of a line segment in a plane is the line which is perpendicular to the line segment and which passes through the midpoint of the line segment. The *Perpendicularity Axiom* states that given a point and a line there is one and only one perpendicular to the line through the point. Thus, we are assured that every line segment has exactly one perpendicular bisector.

If $\overline{pq}$ is a line segment and L is its perpendicular bisector, then, if x is any point on L, $px = xq$, and x is equidistant from p and q. Moreover, if y is any point in the plane which is equidistant from p and q, then y is on L.

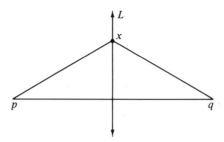

x is on L, the perpendicular bisector of $\overline{pq}$, and x is equidistant from p and q.

Figure 5-5

In other words, we can completely describe the set of points which is the perpendicular bisector of $\overline{pq}$ by the property these points possess of being equidistant from points p and q. In set terminology $L = \{x \mid xp = xq\}$.

Now, given a line segment $\overline{pq}$ we wish to construct its perpendicular bisector.

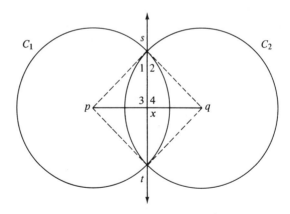

Constructing the perpendicular bisector of $\overline{pq}$

Figure 5-6

1. With center at p, and radius r, where $r > \frac{1}{2}pq$, draw a circle C_1.
2. With center at q and the same radius r, draw a circle C_2.
3. Since $r > \frac{1}{2}pq$, circles C_1 and C_2 will intersect in two points s and t.
4. With straight-edge, draw the line $\overleftrightarrow{st}$.

Now we must show that $\overleftrightarrow{st}$ is the perpendicular bisector of $\overline{pq}$. This means that we must verify that $\overset{\leftrightarrow}{st}$ passes through the midpoint of $\overline{pq}$, and that $\overset{\leftrightarrow}{st}$ is perpendicular to $\overline{pq}$.

To do this, draw the line segments $\overline{ps}, \overline{qs}, \overline{pt}$, and $\overline{qt}$. Then, $\overline{ps} \cong \overline{qs}$ since $ps = qs = r$, and $\overline{pt} \cong \overline{qt}$ for the same reason.

$\triangle spt \cong \triangle sqt$, by the SSS Axiom; therefore $\angle 1 \cong \angle 2$.

It follows from the SAS Axiom that $\triangle psx \cong \triangle qsx$; thus $\angle 3 \cong \angle 4$. Since $\angle 3$ and $\angle 4$ are supplementary, m $\angle 3$ = m $\angle 4 = 90$ and $\overset{\leftrightarrow}{st}$ is perpendicular to $\overline{pq}$; moreover, $\overline{px} \cong \overline{xq}$, hence x is the midpoint of $\overline{pq}$. ∎

This construction really does two things. It gives us a perpendicular to $\overline{pq}$, and also a bisector of the line segment. Thus, if we want to find the midpoint of a segment, we can use the same construction.

Construction 6

Constructing the perpendicular from a given point to a given line

If we want to construct a perpendicular to a line L from a point p, there are two cases to consider: p may be on the line L, or it may not be.

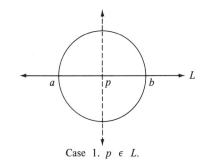

Case 1. $p \in L$.

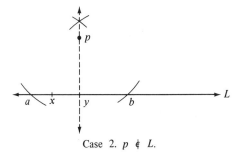

Case 2. $p \notin L$.

Constructing the perpendicular from a given point p to a line L.

Figure 5-7

Case 1 $p \in L$

1. With center at p draw a circle intersecting L at points a and b.

2. Construct the perpendicular bisector of $\overline{ab}$ as in Construction 5. Since p is the midpoint of $\overline{ab}$ (Why?), this will pass through point p.

Case 2 $p \notin L$

1. Choose any point x on L.
2. Draw a circle having center at p and radius r greater than px. This circle will cut L in two points a and b.
3. By Construction 5, find the bisector y of $\overline{ab}$. Line py will then be perpendicular to L.

Construction 7

Constructing a line through a point parallel to a given line
This construction is included here for completeness. However, its proof will be postponed until Chapter 6.

Let L_1 be a line and p any point not on L_1. We wish to construct a line L_2 containing p which is parallel to L_1.

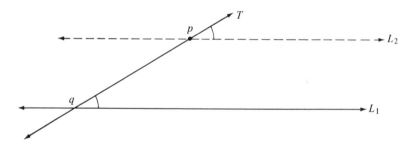

Constructing a line through a point p parallel to a given line L_1

Figure 5-8

1. Draw any line T through p cutting L_1 in a point q.
2. Copy at p an angle congruent to the angle at q.
   The line L_2 will be parallel to L_1.
   An alternate construction will be found in Chapter 6.

Exercise 5.2

The constructions called for in the following problems are to be made using straight-edge and compass only.

1. Draw a line segment $\overline{pq}$. Divide it into four congruent segments.

2. Draw an angle. Construct an angle whose measure is one-fourth that of the given angle.

3. Construct a right angle, and use it to construct an angle whose measure is 45.

4. Given a line segment $\overline{pq}$, construct an isosceles right triangle having two sides which are half the length of $\overline{pq}$.

5. Describe fully how you would copy a triangle using the ASA Axiom.

6. Let L be the perpendicular bisector of $\overline{pq}$ and let x be any point on L. Then L passes through the midpoint r of $\overline{pq}$ and $\angle xrq$ is a right angle. Prove that x is equidistant from p and q, i.e., $xp = xq$.

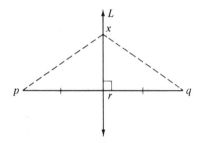

7. Draw $\triangle abc$. Construct the three angle bisectors. Do they intersect in a single point?

8. A *median* of a triangle is a line segment whose end-points are a vertex and the midpoint of the opposite side. Every triangle has three medians. Draw $\triangle abc$. Construct the three medians. Do they intersect in a single point?

9. An *altitude* of a triangle is a line segment from a vertex perpendicular to the line containing the opposite side. Every triangle has three altitudes. Draw $\triangle abc$. Construct its three altitudes. Do the lines containing them intersect in a single point?

10. In Construction 6, Case 2, the statement is made that $\overleftrightarrow{py}$ is perpendicular to L. Prove this statement.

11. Given a line segment $\overline{pq}$, construct a square having pq as the measure of a side.

12. Draw a triangle. Through each vertex construct a line parallel to the opposite side of the triangle. Extend these lines so that they meet in pairs. What do you observe about the three exterior triangles you have constructed?

13. Divide a line segment $\overline{ab}$ into five congruent segments by performing the following construction:

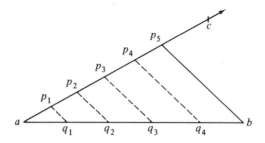

(a) At point a construct any angle.

(b) Along the ray $\overrightarrow{ac}$ mark off with the compass five congruent segments of any convenient length. Label these points p_1, p_2, p_3, p_4, and p_5.

(c) Draw the line segment $\overline{p_5 b}$.

(d) Construct (Construction 7) through p_1, p_2, p_3, and p_4, respectively, lines parallel to $\overline{p_5 b}$ cutting the line segment $\overline{ab}$ in the points q_1, q_2, q_3, and q_4.

These points divide segment $\overline{ab}$ into five congruent segments. Justification of this construction will be found in Chapter 7.

14. Use the construction of problem 13 to divide a line segment into seven congruent line segments.

15. Another way of dividing segment $\overline{ab}$ into any number, say five, of congruent segments uses an ordinary piece of ruled notebook paper.

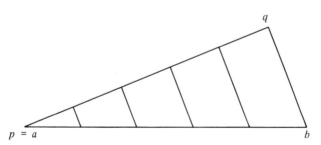

(a) On a piece of lined notebook paper, draw a line segment perpendicular to the lines of the paper so that this segment is divided into five congruent segments by the lines of the paper. Call this segment $\overline{pq}$.

(b) Place the lined paper on the paper on which the segment $\overline{ab}$ is drawn so that the point p lies on the point a.

(c) Keeping p on a, move the lined paper so that the line of the paper containing q passes through point b.

(d)   The lines of the lined paper now intersect $\overline{ab}$ in points which divide it into five congruent segments. Use this construction to divide a line segment into three congruent segments; into seven congruent segments.

3. Other Constructions

We have looked at just a few of the constructions that can be done with ruler and compass. There are many others. One is the Problem of Apollonius. Let C_1, C_2, and C_3 be three circles in the plane. The Problem of Apollonius is to construct with straight-edge and compass all possible circles C which are tangent to C_1, C_2, and C_3. (Two circles are *tangent* if their intersection is a single point.) There are eight solutions to the problem. One of them is illustrated in Figure 5-9.

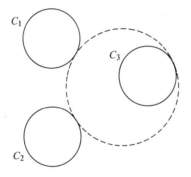

One solution to the Problem of Apollinus

Figure 5-9

Another problem which has attracted the attention of many great mathematicians is that of constructing a regular polygon inside a circle. (A polygon is *regular* if all of its sides are congruent and all of its angles are congruent.)

For some special cases, this construction is quite simple. To construct a square inside a circle, first draw any diameter $\overline{ab}$. Construct the perpendicular bisector of $\overline{ab}$, which will intersect the circle in points c and d. The points a, c, b, and d are the vertices of a square. (Figure 5-10) To construct a regular hexagon in a circle, choose any point a on the circle. Let r be the radius of the circle, and with center at a and radius r,

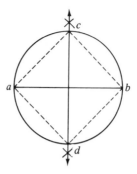

Constructing a square inside a circle

Figure 5-10

draw an arc intersecting the circle in point *b*. Now with center at *b* and the same radius draw an arc intersecting the circle in point *c*. Continue until six points have been so located. These six points are the vertices of a regular hexagon.

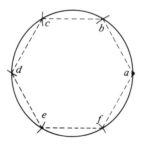

Constructing a regular hexagon inside a circle

Figure 5-11

With these two constructions as a starting point, it is easy to construct regular polygons of 8, 12, 16, or 24 sides in a circle. Many beautiful designs can be drawn using these constructions.

The Greeks knew how to construct regular polygons with 3, 4, 5, 6, 8, 10, and 15 sides with straight-edge and compass alone, and using these figures it is easy to construct regular polygons having twice as many sides. The next step would be to construct regular polygons with 7, 9, 11, or 13 sides. Many attempted this but not one was successful. Finally in 1796 a young man named Gauss was able to prove that the construction is impossible for polygons with 7, 9, 11, or 13 sides. He discovered much more than this. He was able to prove that a straight-edge and compass construction of a regular polygon with an *odd* number of sides is possible if and only if the number of sides is a number of the form $2^{2^n} + 1$ (called a *Fermat prime*) or is a product of *different* Fermat primes.

Thus $3 = 2^{2^0} + 1$, $5 = 2^{2^1} + 1$, $17 = 2^{2^2} + 1$, and $257 = 2^{2^3} + 1$ are Fermat primes, and regular polygons with 3, 5, 17, and 257 sides can be so constructed. Also $15 = 3 \cdot 5$ is a product of two different Fermat primes, and the construction is possible for a polygon with 15 sides. This beautiful discovery, made by Gauss when he was only 19, led him to devote his life to mathematics. He is generally considered to be one of the greatest mathematicians who ever lived.

4. Classical Problems of Antiquity

We have been looking at constructions that can be done with straight-edge and compass. Such a discussion would be incomplete if we did not mention the classical problems of antiquity, three constructions which *cannot* be done with straight-edge and compass alone. These are the problems of the trisection of the angle, the duplication of the cube and the squaring of the circle. The first problem is, given an arbitrary angle, to construct an angle whose measure is one-third that of the given angle. In the second construction, given a line segment representing the edge of a cube, we wish to construct a line segment representing the edge of a cube having twice the volume of the first. Finally, in the third construction, the problem is to construct a square having the same area as a given circle.

An amusing story is told of the origin of the problem of the duplication of the cube. The people of Athens suffered from a plague. They sent a delegation to the oracle at Delos to ask what they might do to appease the gods. They were told that the plague would end if they would double the size of the altar to Apollo, which was cubical in shape. They built a new altar whose edges were twice as long as those of the original one. But this didn't double the size of the altar, it increased the volume eightfold! The demand was not satisfied and the plague continued.

As the name would suggest, these problems date back to the ancient Greeks, and some of the finest mathematicians have tried to solve them and failed. It is only in the past 150 years that the reason for this failure became apparent. Using mathematical methods not available to earlier generations of mathematicians, it has been proved that these constructions cannot be done with straight-edge and compass alone. (For an exceptionally clear exposition of this proof, the reader is referred to Courant and Robbins, *What is Mathematics* (New York: Oxford University Press, Inc., 1941), pp. 117–140.)

The impossibility of these constructions is, of course, a result of the limitations imposed on the tools to be used. If one is permitted to use a marked straight-edge (i.e., a ruler) and compass, for instance, then the problem of the trisection of the angle can be solved.

5. Arithmetic with Straight-edge and Compass

The Greeks were aware of the existence of irrational numbers. They knew that the line segment representing a side of a square and the line segment representing its diagonal were incommensurable, that is, they could not be measured in terms of a common unit. Today we say that a square whose side is one unit in length has a diagonal whose length is $\sqrt{2}$, and $\sqrt{2}$ is an irrational number; it cannot be expressed as the ratio of two integers.

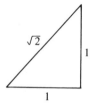

The diagonal of a square and its side are incommensurable.

Figure 5-12

By the symbol "$\sqrt{2}$" we mean that number which when multiplied by itself gives two. The Greeks, lacking our convenient symbols for irrational numbers, treated these numbers geometrically as lengths of line segments. There is certainly no difficulty in thinking of $\sqrt{2}$, for instance, as the length of a diagonal of a certain square.

Sums and Differences

If we think of positive numbers as lengths of line segments, then we can interpret the *sum* of two numbers, say a and b, geometrically. Take two line segments having lengths a and b, respectively, and lay them end to end. Their union is a line segment of length $a + b$. In this way the sum of 1 and $\sqrt{2}$ is no more difficult to find than the sum of 1 and 2.

Similarly, we can construct a segment whose length is the *difference* of two numbers, $a - b$, provided a is larger than b.

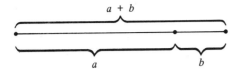

Constructing a line segment whose length is the *sum* of the lengths of two given segments.

Figure 5-13

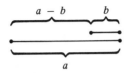

Constructing a line segment whose length is the *difference* of the lengths of two given segments.

Figure 5-14

Products

Now suppose that we wish to construct a line segment whose length is the *product* of the lengths of two segments $\overline{ab}$ and $\overline{ad}$. To do this, start with any angle and label its vertex a. Along one side, copy the segment $\overline{ab}$ and along the other the segment $\overline{ad}$.

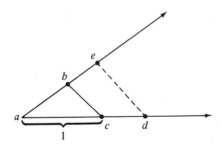

The length of line segment $\overline{ae}$ is the *product* of the lengths of line segments $\overline{ab}$ and $\overline{ad}$.

Figure 5-15

Now construct segment $\overline{ac}$ having length one unit, where c lies on the ray $\overrightarrow{ad}$. With a straight-edge draw $\overline{cb}$. Now construct (Construction 7) a line through d parallel to $\overline{bc}$, cutting $\overrightarrow{ab}$ at e. The segment $\overline{ae}$ has the desired length.

The two triangles acb and ade are called similar triangles since they have the same shape, and it will be shown in Chapter 7 that if two triangles are similar, then the ratios of the lengths of their corresponding sides are equal.

Thus, we have

$$\frac{ae}{ab} = \frac{ad}{ac}$$

and since $ac = 1$,

$$ae = (ad) \times (ab)$$

We have constructed a line segment $\overline{ae}$ whose length is the product of the lengths of segments $\overline{ad}$ and $\overline{ab}$.

Quotients

The construction of a segment whose length is the *quotient* of the lengths of two segments is similar. To find $(ab)/(ad)$, again take any angle a and copy the segments $\overline{ab}$ and $\overline{ad}$ along its sides. Draw segment $\overline{bd}$. Draw a segment $\overline{ac}$ having length one unit on ray $\overrightarrow{ad}$ and construct (Construction 7) a line through c parallel to $\overline{bd}$ cutting the ray $\overrightarrow{ab}$ at e. The segment $\overline{ae}$ has the desired length.

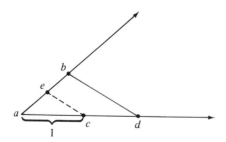

The length of line segment $\overline{ae}$ is the *quotient* of the lengths of segments $\overline{ab}$ and $\overline{ad}$.

Figure 5-16

As before, $\triangle ace$ is similar to $\triangle adb$, hence

$$\frac{ae}{ab} = \frac{ac}{ad}$$

and since $ac = 1$,

$$ae = \frac{ab}{ad}$$

Square Roots

We have already constructed a segment having length $\sqrt{2}$ by drawing a diagonal of a square whose side measures one unit. We can construct a segment whose length is the square root of the length of *any* line segment using only straight-edge and compass. Suppose we wish to construct a segment whose length is the square root of the length of segment $\overline{ab}$.

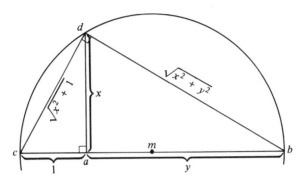

The length of line segment $\overline{ad}$ is the square root of the length of segment $\overline{ab}$.

Figure 5-17

First, we construct segment $\overline{ca}$ having length one unit. Next, find the midpoint of $\overline{cb}$, call it m (Construction 5). Now with center at m, draw a circle having radius mb and construct a perpendicular to $\overline{cb}$ at a (Construction 6), cutting the circle in the point d. The segment $\overline{ad}$ will have the desired length.

To justify this construction we use the Pythagorean Theorem (Chapter 7) which states that in a right triangle the square of the length of the hypotenuse (the side opposite the right angle) is equal to the sum of the squares of the lengths of the other two sides. For convenience, call the length of $\overline{ad}$, x, and the length of $\overline{ab}$, y. Then, since $\triangle adc$ is a right triangle,

$$x^2 + 1 = (cd)^2$$

or

$$cd = \sqrt{x^2 + 1}$$

Now, $\triangle abd$ is also a right triangle, so

$$x^2 + y^2 = (db)^2$$

or

$$db = \sqrt{x^2 + y^2}$$

It turns out that $\angle cdb$ is also a right angle, hence

$$(cd)^2 + (db)^2 = (cb)^2$$

or

$$(x^2 + 1) + (x^2 + y^2) = (y + 1)^2$$

Then,

$$2x^2 + y^2 + 1 = y^2 + 2y + 1$$
$$2x^2 = 2y$$
$$x^2 = y$$

and

$$x = \sqrt{y}$$

One might think that since we have been able to construct square roots we could proceed next to construct cube roots. Oddly enough, it turns out that it is impossible (except for special cases) to construct a line segment whose length is the cube root of the length of a given line segment using straight-edge and compass alone. In fact, this is a crucial point in proving that it is impossible to trisect an angle using only these tools. We can, however, construct a *fourth* root. Since a fourth root is a square root of a square root ($\sqrt[4]{2} = \sqrt{\sqrt{2}}$), this involves only performing the preceding construction twice.

Exercise 5.5

1. Construct a regular polygon with eight sides inside a circle.

2. Construct a regular polygon with twelve sides inside a circle.

3. Construct an equilateral triangle inside a circle.

4. Go through the construction of the product of $\overline{ab}$ and $\overline{ad}$ in the text taking $\overline{ad}$ to be one unit in length. What is the product $\overline{ae}$ in this case?

5. Go through the construction of the product of $\overline{ab}$ and $\overline{ad}$ in the text taking $\overline{ab}$ to be one unit in length. Explain why the product $\overline{ae}$ is the same length as $\overline{ad}$.

Perform the following constructions taking a unit to be any convenient fixed length, say one inch.

6. Construct a segment of length $\sqrt{2}$ using the method given in the text. Construct a segment of length $\sqrt{2}$ as the diagonal of a square having a side measuring one unit. Compare the lengths of the two segments.

7. Construct line segments having the following lengths:

    (a) $1 + \sqrt{2}$ (b) $\sqrt{2} - 1$

    (c) $\sqrt{2} \cdot \sqrt{2}$ (d) $\sqrt{2} \div \sqrt{2}$

8. Construct line segments having lengths:

   (a) $\sqrt{3}$ (b) $\sqrt[4]{3}$

9. The following figure illustrates a method of finding the successive square roots $\sqrt{2}, \sqrt{3}, \sqrt{4}, \ldots$

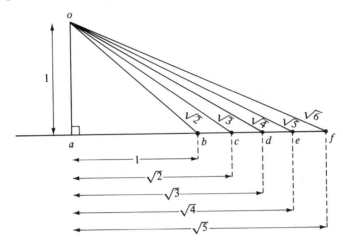

   (a) Construct $\overline{oa} \perp \overleftrightarrow{ab}$, and take $\overline{oa}$ and $\overline{ab}$ to have measures one unit.
   (b) Draw $\overline{ob}$.
   (c) Construct $\overline{ac}$ congruent to $\overline{ob}$.
   (d) Draw $\overline{oc}$.
   (e) Construct $\overline{ad}$ congruent to $\overline{oc}$.
   ... etc.

   Explain why $ac = \sqrt{2}$; $ad = \sqrt{3}$, $ae = \sqrt{4}$, $af = \sqrt{5}$, etc.

New Terms Found in This Chapter

Term	Section
bisector of an angle	2
perpendicular bisector of a line segment	2
tangent circles	3
regular polygon	3
median	Exercise 5.2
altitude	Exercise 5.2

6: Parallels and Parallelograms

1. Tests for Parallelism

In defining the parallelism relation between two lines, we have said that two lines are parallel if they lie in the same plane and do not intersect. If L_1 and L_2 are parallel lines, then we write $L_1 \parallel L_2$. Just as we defined perpendicularity between two rays or two line segments, so we can say that two rays or two line segments are parallel if the lines containing them are parallel.

This definition of parallel lines, while descriptive, is not really very useful. Two line segments in a plane might appear to be parallel, but since the lines determined by them are infinite in extent, we might not be able to tell whether or not these lines intersect. Accordingly, we need to develop some tests other than the definition for parallelism. One such test is given by the following theorem:

THEOREM 1 If two lines lie in the same plane and are perpendicular to the same line, then they are parallel.

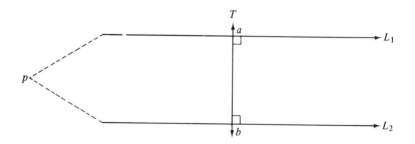

If L_1 and L_2 are both perpendicular to T, then $L_1 \parallel L_2$.

Figure 6-1

112

Proof: To prove this theorem we will make use of the Perpendicularity Axiom (Chapter 5, Section 2). Recall that the Perpendicularity Axiom states that given a line L and a point p not on L one and only one line may be drawn through p perpendicular to L.

Let L_1, L_2, and T be three lines lying in a plane, such that $L_1 \perp T$ and $L_2 \perp T$. We want to show that $L_1 \parallel L_2$. The proof will be indirect. Suppose L_1 is not parallel to L_2. Then $L_1 \cap L_2$ contains exactly one point p. Thus, there are two lines, L_1 and L_2, through p which are perpendicular to T. This contradicts the Perpendicularity Axiom, hence the assumption that L_1 and L_2 intersect must be false and $L_1 \parallel L_2$. ∎

Since this theorem depends heavily on the Perpendicularity Axiom for its proof, it is not surprising to find that in a model for which the Axiom does not hold, the theorem is not true. Recall the model of the sphere in which lines are great circles. If we take the point p to be the North Pole and the line L to be the equator, then infinitely many lines may be drawn through p perpendicular to L. These are the lines of longitude. These lines are all perpendicular to the equator, but they are not parallel as they all meet at the North Pole.

In Theorem 1, lines L_1 and L_2 were both perpendicular to line T. Now we will develop a test for parallelism which is more general than Theorem 1, but first we need to define some terms.

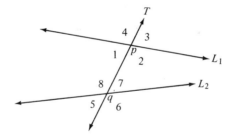

L_1 and L_2 are cut by the transversal T. $\angle 1$ and $\angle 7$ are a pair of alternate interior angles. $\angle 3$ and $\angle 7$ are a pair of corresponding angles.

Figure 6-2

If L_1, L_2, and T are three lines in a plane, and if T intersects L_1 and L_2 in two different points p and q, then T is a *transversal* of L_1 and L_2. We commonly say that L_1 and L_2 are cut by the transversal T.

When two lines are cut by a transversal, a number of angles are formed. Certain pairs of these have special names.

In Figure 6-2 $\angle 1$ and $\angle 7$ are called *alternate interior angles*, as are $\angle 2$ and $\angle 8$.

From an earlier discussion, we recognize a number of pairs of vertical angles in this figure. (How many?)

If $\angle 1$ and $\angle 3$ are vertical angles and $\angle 1$ and $\angle 7$ are alternate interior angles, then we will call $\angle 3$ and $\angle 7$ *corresponding angles.* Extending this to other pairs of angles we see that in the figure $\angle 2$ and $\angle 6$ are corresponding angles also, and in general, if $\angle a$ and $\angle b$ are vertical angles and $\angle a$ and $\angle c$ are alternate interior angles, then $\angle b$ and $\angle c$ are corresponding angles.

It is not hard to see that if one pair of corresponding angles is congruent, then every other pair of corresponding angles is congruent, and alternate interior angles are congruent as well. (See Exercise 6.1, problems 2 through 5.)

Now suppose that a pair of corresponding angles is congruent, say $\angle 3 \cong \angle 7$. (Figure 6-3) Then, $\angle 1 \cong \angle 5$, since they are corresponding angles also. If we turn the picture upside down, so that L_2 is above L_1, then $\angle 5$ is in the position $\angle 3$ was in the first figure and $\angle 1$ lies in the position formerly occupied by $\angle 7$.

This suggests that the part of the picture to the right of the transversal T and that to its left look alike. It seems reasonable that if L_1 and L_2

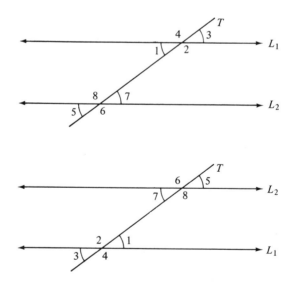

If $\angle 3 \cong \angle 7$ then $\angle 1 \cong \angle 5$, and the picture must look the same if it is turned upside down.

Figure 6-3

meet somewhere to the right, then they must meet somewhere to the left as well. This, however, is clearly impossible, since two lines can meet in at most one point. (Theorem 1, Chapter 2) We are thus led to the conjecture that the two lines do not meet at all, and are therefore parallel.

In order to prove this conjecture, we will need first to define an exterior angle of a triangle and prove a preliminary theorem. (A theorem whose chief purpose is to make the proof of a main theorem simpler is sometimes called a *lemma*.)

If we have a triangle *abc* and we extend one side, say $\overline{ab}$ to form a ray $\overrightarrow{ad}$, then $\angle cbd$ is called an *exterior angle* of the triangle. (Figure 6-4) Angles *a* and *c* are called *opposite interior angles* to this exterior angle. Our preliminary theorem, or lemma, involves the measure of these angles.

Lemma The measure of an exterior angle of a triangle is greater than the measure of either of the opposite interior angles.

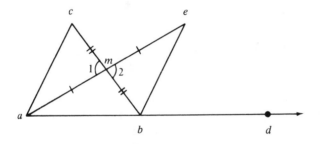

$\angle cbd$ is an exterior angle of $\triangle abc$. Its measure is larger than that of either $\angle a$ or $\angle c$.

Figure 6-4

Proof: First we need to construct some auxiliary lines. Find *m*, the midpoint of $\overline{cb}$ and draw $\overline{ae}$ so that $\overline{am} \cong \overline{me}$. Draw $\overline{eb}$.

Now $\overline{cm} \cong \overline{mb}$ by construction, $\overline{am} \cong \overline{me}$ by construction, and $\angle 1 \cong \angle 2$, since they are vertical angles. Thus, $\triangle amc \cong \triangle emb$ by the SAS Axiom, and consequently $\angle cbe \cong \angle c$.

But *e* is in the interior of the exterior angle, $\angle cbd$*; hence by the properties of measure of angles, m $\angle cbe <$ m $\angle cbd$. We conclude that m $\angle c <$ m $\angle cbd$. Similarly, we can prove that m $\angle a <$ m $\angle cbd$. ∎

Now we use this lemma to prove our conjecture.

---

*This proof depends on the picture, in that we are assuming *e* is in the interior of $\angle cbd$. In a rigorous development, we would have to prove this fact.

THEOREM 2 If two lines in a plane are cut by a transversal so that a pair of corresponding angles are congruent, then the two lines are parallel.

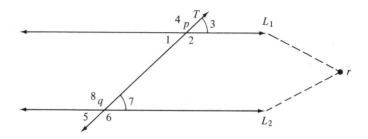

If a pair of corresponding angles are congruent, then $L_1 \parallel L_2$.

Figure 6-5

Proof: Let L_1 and L_2 be two lines in a plane cut by the transversal T in points p and q, respectively, and assume that $\angle 3 \cong \angle 7$. We wish to prove that L_1 and L_2 are parallel. Our proof will be by contradiction. We will assume that the lines are not parallel and show that this leads to a contradiction of the lemma we have just proved.

Suppose L_1 and L_2 are not parallel. Then they meet in some point r, which we will assume is to the right of T. Then, pqr is a triangle and $\angle 8$ is an exterior angle of this triangle. But $\angle 8 \cong \angle 2$ (Why?) and $\angle 2$ is an opposite interior angle to $\angle 8$. This contradicts our lemma, which says the measure of an exterior angle must be *greater than* the measure of an opposite interior angle. We conclude then that $L_1 \parallel L_2$. ∎

If r is to the left of T, the proof is similar. Only the names of the angles have to be changed.

Corollary If two lines in a plane are cut by a transversal so that a pair of alternate interior angles are congruent, then the two lines are parallel.

Exercise 6.1

1. Name all pairs of corresponding angles in the figure following. Name all pairs of alternate interior angles.

2. In the figure following, assume $\angle 3 \cong \angle 7$. Prove that all of the other pairs of corresponding angles are congruent.

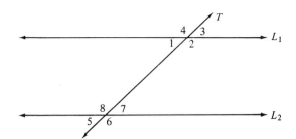

3. Assume $\angle 3 \cong \angle 7$. Prove that both pairs of alternate interior angles are congruent.

4. Assume $\angle 1 \cong \angle 7$. Prove that $\angle 2 \cong \angle 8$.

5. Assume $\angle 1 \cong \angle 7$. Prove that all pairs of corresponding angles are congruent.

6. Give reasons for each step in the following proof:

Given: $\overline{ca} \cong \overline{cb}$

m the midpoint of $\overline{ab}$

$\overline{cm} \perp \overline{de}$

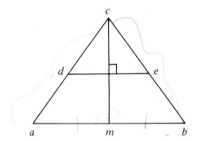

Prove: $\overline{de} \parallel \overline{ab}$

Proof: Reasons
1. $\overline{ca} \cong \overline{cb}$ 1.
2. $\overline{am} \cong \overline{mb}$ 2.
3. $\overline{cm} \cong \overline{cm}$ 3.
4. $\triangle acm \cong \triangle bcm$ 4.
5. $\angle amc \cong \angle bmc$ 5.
6. $\overline{cm} \perp \overline{ab}$ 6.
7. $\overline{cm} \perp \overline{de}$ 7.
8. $\overline{ab} \parallel \overline{de}$ 8.

7.

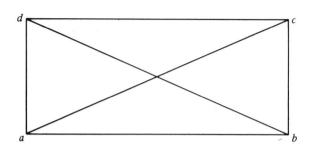

In the figure above, name all transversals of $\overleftrightarrow{dc}$ and $\overleftrightarrow{ab}$. Name all transversals of $\overleftrightarrow{db}$ and $\overleftrightarrow{ac}$; of $\overleftrightarrow{ad}$ and $\overleftrightarrow{ac}$. Is $\overleftrightarrow{ab}$ a transversal of $\overleftrightarrow{ad}$ and $\overleftrightarrow{ac}$? Why or why not?

8. Complete the proof of the lemma by giving reasons for each step in the following proof.

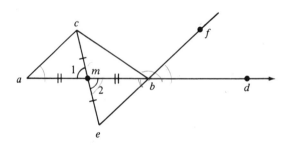

Construct $\overline{ce}$ so that $\overline{cm} \cong \overline{me}$, and m is the midpoint of $\overline{ab}$.

Proof:	Reasons
1. $\overline{am} \cong \overline{mb}$	1.
2. $\overline{cm} \cong \overline{me}$	2.
3. $\angle 1 \cong \angle 2$	3.
4. $\triangle amc \cong \triangle bme$	4.
5. $\angle a \cong \angle mbe$	5.
6. $\angle fbd \cong \angle mbe$	6.
7. m $\angle cbd >$ m $\angle fbd$	7.
8. m $\angle cbd >$ m $\angle a$	8.

9. Complete the proof of Theorem 2 for the case in which L_1 and L_2 meet in a point r to the *left* of the transversal T.

10. Prove the Corollary to Theorem 2.

2. The Parallel Axiom

We can use Theorem 1 to construct a line parallel to a given line through a given point not on the line.

Construction 8

Let L_1 be a line and p a point not on L_1.

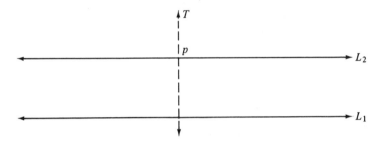

Constructing a line through p parallel to L_1

Figure 6-6

By construction 6 (Chapter 5) construct a perpendicular T to L_1 through point p. Then using Construction 6 again, construct a perpendicular L_2 to T through p. By Theorem 1, $L_1 \parallel L_2$.

This construction assures us that given a line L and a point p not on L, there exists *at least one* line parallel to L. Note that to establish this fact we needed the Perpendicularity Axiom which was used in proving Theorem 1. Strangely enough, we cannot prove that there is *only one* parallel to L through p using only the axioms stated thus far. We must state the property as an axiom, the Parallel Axiom.

The Parallel Axiom   Given a line L and a point p not on L there is one and only one line through p parallel to L.

This axiom is often called the *Euclidean Parallel Axiom*. Actually, this particular form of the axiom is not due to Euclid, but to a mathematician named Playfair, and it is also called Playfair's Axiom. Euclid's version of this axiom goes like this:

"If a straight line falling on two straight lines makes the interior angles on the same side less than two right angles, the two straight lines, if produced indefinitely, meet on that side on which are the angles less than two right angles."

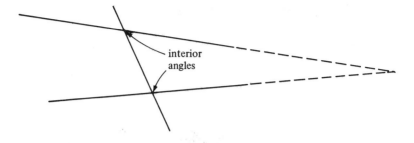

Euclid's Parallel Axiom states that if the sum of the measures of the two interior angles is less than 180, then the lines, if produced indefinitely, will meet on that side.

Figure 6-7

The two axioms are equivalent. By this we mean that if we assume Euclid's statement as an axiom (together with his other axioms), then Playfair's Axiom can be proved as a theorem. Conversely, if we assume Playfair's Axiom and the axioms of Euclid (with the exception of the parallel axiom), then Euclid's Parallel Axiom can be proved as a theorem. Since they are equivalent, we can use them interchangeably, and as Playfair's Axiom is simpler in form, it is the one we commonly use.

Using the Parallel Axiom, we can prove that the converse of Theorem 2 (if two parallel lines are cut by a transversal, then the corresponding angles are congruent) is true. (Figure 6-8)

THEOREM 3   If parallel lines are cut by a transversal, then the corresponding angles are congruent.

Proof: Let L_1 and L_2 be parallel lines cut by a transversal T, and suppose that $\angle 1 \not\cong \angle 2$. Then by Construction 2 (Chapter 5), we can construct a line L_3 through p so that $\angle spr \cong \angle 2$. Then, since a pair of corresponding angles is congruent, by Theorem 2, $L_3 \parallel L_2$. But $L_3 \neq L_1$ since r is on L_3 but not on L_1. Thus, we have two distinct lines through p, both of which are parallel to L_2. This is impossible by the Parallel Axiom. Since we have reached a contradiction, we conclude that $\angle 1 \cong \angle 2$. ∎

Corollary   If parallel lines are cut by a transversal, then the alternate interior angles are congruent.

We can use these results to prove the following theorems:

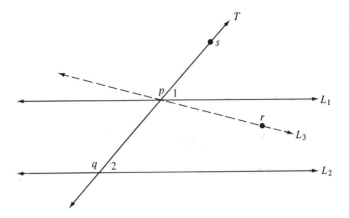

If $L_1 \, /\!/ \, L_2$, then corresponding angles are congruent.

Figure 6-8

THEOREM 4   If a transversal is perpendicular to one of two parallel lines, then it is perpendicular to the other as well.

THEOREM 5   If two different lines in a plane are parallel to a third line, then they are parallel to each other.

The proof of these theorems is left for the student.

A Greek mathematician named Eratosthenes who lived two-and-one-half centuries before Christ used these ideas to calculate the circumference of the earth with remarkable accuracy. He found that at noon on the longest day of the year the sun is directly overhead at the city of Scyene in Egypt; a rod stuck upright in the ground there casts no shadow. At the same moment an upright rod at Alexandria casts a short shadow. Eratosthenes knew that the earth was a sphere and that the rays of the sun are nearly parallel to each other when they strike the earth. Using these facts he probably drew a picture somewhat like Figure 6-9.

Since the line from the center of the earth to Alexandria is a transversal cutting the parallel lines which are the sun's rays, the angle abc must be congruent to the angle θ at the center of the earth. The $\angle abc$ was measured and found to be $7\frac{1}{2}°$, or one-fiftieth of $360°$. Erathostenes concluded that the distance from Scyene to Alexandria must be one-fiftieth of the circumference of the earth. Since the distance between the two cities was known, he calculated that the circumference of the earth was approximately 25,000 miles. Today the figure is known to be approximately 24,907. Eratosthenes' estimate was incorrect by only 93 miles, or about .4%!

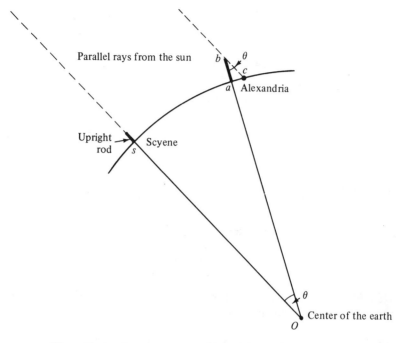

How Eratosthenes measured the circumference of the earth. The lengths of the rods and the distance from Scyene to Alexandria are, of course, greatly exaggerated in this sketch.

Figure 6-9

Exercise 6.2

1. Describe in detail how you would construct a line parallel to a given line L through a point p not on L by using Theorem 2. By using the Corollary to Theorem 2.

2. Prove the Corollary to Theorem 3.

3. Let L_1 and L_2 be parallel lines cut by a transversal T. Show that $\angle 1$ and $\angle 3$ and $\angle 2$ and $\angle 4$ are supplementary angles. (See the figure following.)

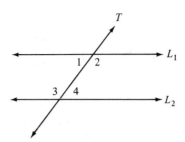

4. Prove Theorem 4. Let L_1, L_2, and T be three lines in a plane, $L_1 \parallel L_2$, $L_1 \perp T$. Prove that $L_2 \perp T$.

5. Prove Theorem 5. Given three distinct lines L_1, L_2, L_3 in a plane, $L_1 \parallel L_2, L_2 \parallel L_3$, prove that $L_1 \parallel L_3$. (*Hint:* Draw a transversal.)

6. Consider the relation "is parallel to" between lines in the plane. Is the relation reflexive? Symmetric? Transitive? Explain.

7.

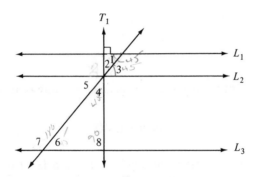

In the figure $L_1 \parallel L_2, L_2 \parallel L_3$, $T_1 \perp L_1$, m $\angle 1 = 45$. Find the measure of angles 2, 3, 4, 5, 6, 7, 8. Give reasons for your statements.

8.

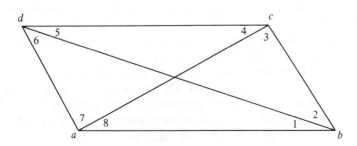

In the preceding figure, which line segments (if any) may be shown to be parallel by each of the statements below?

(a) $\angle 1 \cong \angle 5$ (b) $\angle 2 \cong \angle 5$

(c) $\angle 4 \cong \angle 8$ (d) $\angle 3 \cong \angle 7$

3. The Angle Sum Theorem

We can now use the Parallel Axiom to prove the well-known theorem that the sum of the measures of the angles of any triangle is 180.

THEOREM 6   In any triangle abc, m $\angle a$ + m $\angle b$ + m $\angle c$ = 180.

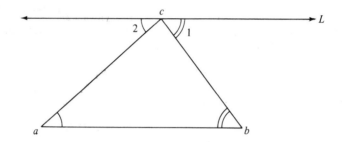

The sum of the measures of the angles of any triangle is 180.

Figure 6-10

Proof: First we construct through c a line L parallel to $\overline{ab}$. Clearly, m $\angle 2$ + m $\angle c$ + m $\angle 1$ = 180. By Theorem 3, since $L \parallel \overline{ab}$, $\angle 2 \cong \angle a$ and $\angle 1 \cong \angle b$, hence m $\angle 2$ = m $\angle a$ and m $\angle 1$ = m $\angle b$. Therefore, m $\angle a$ + m $\angle b$ + m $\angle c$ = 180. ∎

Note that to prove this theorem we used Theorem 3, and to prove Theorem 3 we needed the Parallel Axiom. Theorem 6 thus depends on the Parallel Axiom.

A convincing demonstration of this theorem can be given by the following paper-folding device. (Figure 6-11)

Cut a triangle abc from paper. Fold along the dotted lines.

The vertices of the triangles then will meet as in the figure so that it can be seen that the sum of the measures of these angles is 180.

The student is cautioned that this is *not* a proof. Paper cutting and folding devices can be used to demonstrate many propositions, including some that are not true.

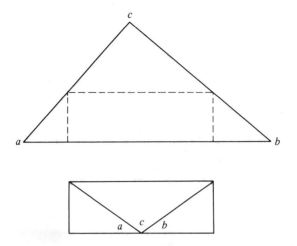

Demonstrating the angle sum theorem by paper-folding

Figure 6-11

As an example of this, consider the following bit of hocus-pocus. The big square in Figure 6-12 contains 64 small squares, eight on a side. If this is cut apart on the lines indicated and rearranged, it forms a rectangle which has 5 squares on one side, 13 on the other, or 65 squares in all. Where did the extra square come from?

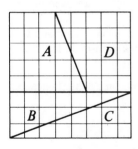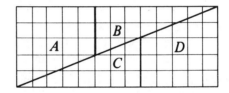

In cutting the square apart and reassembling it we gain one square unit of area.

Figure 6-12

The explanation of this apparent contradiction is that the pieces only *appear* to fit together in the second picture. The edges of pieces *A*, *B*, *C*, and *D* do not actually coincide along the diagonal, but form a parallelogram *pqrs* (Figure 6-13) whose area is the extra square unit. The angle at *p* is so small that the gap is never noticed in cutting and rearranging the pieces.

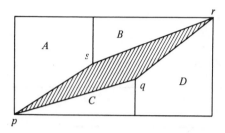

The extra square unit is found hiding in the crack.

Figure 6-13

Recall that in Chapter 4 we stated, without proof, that in the ASA Axiom it was not necessary that the given side be included between the two angles. We can now justify this assertion by use of the following theorem:

THEOREM 7 If two angles of a triangle are congruent respectively to two angles of a second triangle, then the third angles are congruent as well.

The proof of this theorem is left to the student as an exercise.

AAS THEOREM If two angles and a non-included side of one triangle are congruent respectively to two angles and a non-included side of a second triangle, then the triangles are congruent.

Proof: By Theorem 7 the third angles are congruent, hence by the ASA Axiom, the two triangles are congruent. ▮

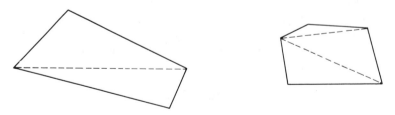

Finding the sum of the measures of the angles of a polygon

Figure 6-14

Now consider the problem of finding the sum of the measures of the angles of a convex quadrilateral.

From any vertex, we draw a diagonal. This gives us two triangles. It is easy to see that the sum of the measures of the angles of the two triangles is the same as the sum of the measures of the angles of the quadrilateral. Therefore, we conclude that the sum of the measures of the angles of a convex quadrilateral is 2 · 180.

In the same way, from one vertex of a convex pentagon we can draw two diagonals. This gives us three triangles, the sum of the measures of whose angles is 3 · 180. We conclude that the sum of the measures of the angles of a convex pentagon is 3 · 180. The student should have no trouble in extending this to a statement about the sum of the measures of the angles of a convex polygon of n sides (Problem 15).

Exercise 6.3

1. If the measures of two angles of a triangle are as follows, what is the measure of the third angle?

   (a) 10 and 40   (b) 100 and 49
   (c) r and r (d) 60 and 60
   (e) $(45 + x)$ and $(45 - x)$ (f) 30 and 60

2. In the figure find the measures of angles 1, 2, 3, 4, and 5.

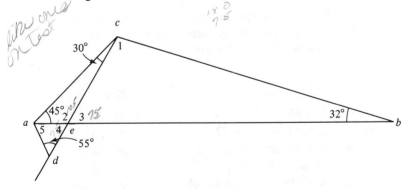

3. Prove that the measure of an exterior angle of a triangle is equal to the sum of the measures of the two opposite interior angles.

4. Prove Theorem 7. That is, given $\triangle abc$ and $\triangle def$, with $\angle a \cong \angle d$ and $\angle b \cong \angle e$, show that $\angle c \cong \angle f$. (See the figure following.)

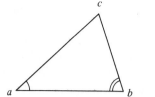

5. Given △*abc*, how many exterior angles does it have? Label each exterior angle and express the measure of each as the sum of the measures of the two opposite interior angles.

6. Show that if △*abc* is a right triangle then its acute angles are complementary.

7. Show that the measure of each of the angles of an equilateral triangle is 60.

8. The measure of one angle of a triangle is twice the measure of a second angle. The measure of the third angle is three times the measure of the second angle. Find the measure of each angle.

9. The measure of a base angle of an isosceles triangle is twice the measure of a second angle. Find the measure of each angle of the triangle.

10. Suppose that the measure of one angle of an isosceles triangle is 50. Find the measures of the other two angles. (*Hint:* There are 2 answers to this problem.)

11. In the following table the measures of various parts of △*abc* are given and you are to find the measure of some other part. In each case, give reasons for your answer.

	Given	*Find*
(a)	m ∠*b* = 60; m ∠*c* = 40	m ∠*a*
(b)	m ∠*a* = 52; m ∠*c* = 37	m ∠*b*
(c)	m ∠*b* = 40; *ab* = 2; *ac* = 2	m ∠*a*
(d)	*ab* = 3; *ac* = 3; *bc* = 3	m ∠*c*
(e)	m ∠*a* = 100; m ∠*c* = 40; *ab* = 4	*ac*

12. In the figure following $\overline{ab}$ // $\overline{cd}$ and m ∠10 = 70, m ∠8 = 30, m ∠9 = 95. Find the measures of the other numbered angles. Give reasons for your answers.

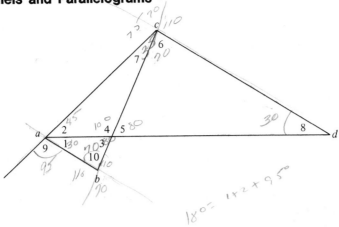

13.

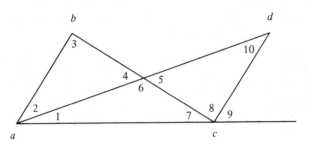

In the figure above $\overline{ab}$ $\parallel$ $\overline{cd}$, m $\angle 10 = 45$, m $\angle 8 = 80$, m $\angle 9 = 75$. Find the measures of the other numbered angles. Give reasons for your answers.

14. The sum of the measures of the angles of a convex quadrilateral is $2 \cdot 180$. Is this statement true if the quadrilateral is not convex? Illustrate by a specific numerical example.

15. Given a convex polygon, from any one vertex we can draw a number of diagonals and thus form a number of triangles. We can use this to find the sum of the measures of the angles of the polygon. Complete the following table.

Number of Sides	Number of Triangles Formed	Sum of the Measures of the Angles
3		
4		
5		
.		
.		
n		

16. A convex polygon is *regular* if all of its sides are congruent and all of its angles are congruent. What is the measure of each angle of a regular polygon of 3 sides? Of 4 sides? Of 5 sides? Of *n* sides?

17. Find the number of sides of a convex polygon if the sum of the measures of its angles is

    (a) 2700   (b) 1260
    (c) 1980   (d) 4140

18. Could the sum of the measures of the angles of a convex polygon be 920? Why?

19. How many sides does a convex polygon have if each of its angles has a measure of 150? Of 108?

20. What is the largest number of obtuse angles that a convex quadrilateral may have? What is the largest number of acute angles that a convex quadrilateral may have? Explain.

21. If we extend the sides of any convex polygon as in the figure, the sum of the measures of the exterior angles indicated will always be 360, no matter how many sides the polygon may have. This may be demonstrated by cutting out the angles and fitting them together. Demonstrate this statement for a convex polygon of 6 sides, 7 sides, 10 sides.

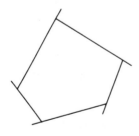

22. Suppose that you wish to cover a floor with tiles and these tiles are to be congruent regular polygons.

    (a) Show that the measure of each angle of a regular polygon is $(n - 2)180/n$, where n is the number of sides of the polygon.
    (b) These tiles are to be arranged so that each vertex of one touches a vertex of the tiles adjacent to it. Show that the number of polygons which touch at each vertex is $2n/(n - 2) = 2 + 4/(n - 2)$.
    (c) Show that, consequently, $n = 3, 4,$ or 6, and the floor can be tiled only with regular triangular, square, or hexagonal tiles.
    (d) Sketch these three patterns.

23. In the floor tiling problem, suppose that the polygons are to be placed so that each vertex of one touches a *side* of one of those polygons adjacent to it.

  (a) Show that the number of polygons meeting at each vertex must be $1 + n/(n-2) = 2 + 2/(n-2)$.
  (b) Show that, consequently, we must have $n = 3$ or 4.
  (c) Sketch these two patterns.

4. Parallelograms

To draw parallel lines draftsmen use an instrument called parallel rulers that looks something like Figure 6-15.

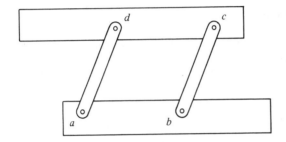

Parallel Rulers

Figure 6-15

It consists of two straight-edges $\overline{ab}$ and $\overline{dc}$ connected by two upright pieces in such a way that the angles are not fixed but pivot freely. The distance ab is the same as distance dc, and distance ad is the same as bc. To draw a line through a point p parallel to line L, place one of the rulers, say $\overline{ab}$ on the line, then shift the upper part of the instrument until the upper ruler $\overline{dc}$ touches the point p. Line segments $\overline{ab}$ and $\overline{dc}$ will always be parallel.

The figure $abcd$ is an example of a special kind of quadrilateral called a parallelogram. A quadrilateral is called a *parallelogram* if both pairs of opposite sides are parallel. The symbol $\square$ $abcd$ is used to denote the parallelogram with vertices $a, b, c,$ and d.

A parallelogram has many interesting properties that quadrilaterals in general do not have. We have seen, for example, that the opposite sides of the figure formed by the parallel rulers have the same measure. Most of these properties can be deduced from the fact that either diagonal of a parallelogram divides the parallelogram into two congruent triangles.

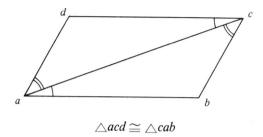

$$\triangle acd \cong \triangle cab$$

Figure 6-16

Using this congruence, the following theorems are easy to prove.

THEOREM 8   Opposite sides of a parallelogram are congruent.

THEOREM 9   Opposite angles of a parallelogram are congruent.

Proof of these theorems is left for the student.

THEOREM 10   Consecutive angles of a parallelogram are supplementary.

Proof: For $\square abcd$, m $\angle a$ + m $\angle b$ + m $\angle c$ + m $\angle d = 2 \cdot 180$. Since by Theorem 9, m $\angle a = m \angle c$ and m $\angle b = m \angle d$, we have 2m $\angle a + 2$m $\angle b = 2 \cdot 180$, hence m $\angle a$ + m $\angle b = 180$.

From this it is easy to show that *any* 2 consecutive angles of $\square abcd$ are supplementary. ∎

The preceding theorems give us some of the properties of parallelograms. The following theorem gives us a way, other than the definition, of telling when a quadrilateral is a parallelogram.

THEOREM 11   A quadrilateral having both pairs of opposite sides congruent is a parallelogram. (Figure 6-16)

The student should complete the proof by giving reasons for each statement. (Exercise 6.5, problem 6)

Proof:	Reasons
1. $\overline{ab} \cong \overline{cd}$ and $\overline{ad} \cong \overline{bc}$	1.
2. $\triangle abc \cong \triangle cda$	2.
3. $\angle dca \cong \angle cab$	3.
4. $\overline{dc} \parallel \overline{ab}$	4.
5. $\angle dac \cong \angle acb$	5.
6. $\overline{ad} \parallel \overline{bc}$	6.
7. The quadrilateral *abcd* is a parallelogram.	7.

Sometimes parallel lines are defined to be two lines which are everywhere equidistant. To examine this statement more closely, we need to decide what is meant by the distance between two parallel lines.

First of all, it is perfectly clear what is meant by the distance between two points. Now we must decide how we will measure the distance from a point to a line.

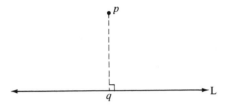

The distance from the point p to the line L is the number pq, where $\overline{pq} \perp L$.

Figure 6-17

Let us agree that the *distance from a point p to a line L* will be the perpendicular distance, i.e., in Figure 6-17 the measure of pq, where $\overline{pq} \perp L$. Now with these ideas in mind we can define the *distance between two parallel lines* to be the distance from any point on one line to the other line.

Thus, the statement "parallel lines are everywhere equidistant" means that if $L_1 \parallel L_2$ and if p and q are any two points on L_1 (Figure 6-18), then the distance from p to L_2 is the same as the distance from q to L_2.

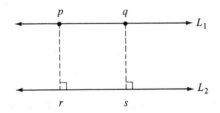

Parallel lines are everywhere equidistant.

Figure 6-18

This statement is not difficult to prove. Since $\overline{pr}$ and $\overline{qs}$ are both perpendicular to L_2, then $\overline{pr} \parallel \overline{qs}$ by Theorem 1. Moreover, $L_1 \parallel L_2$. Thus the quadrilateral $prsq$ is a parallelogram and by Theorem 8 $\overline{pr} \cong \overline{qs}$.

5.   Classification of Quadrilaterals

A parallelogram is a certain kind of quadrilateral. Other special types of quadrilaterals are the trapezoid, the rhombus, the rectangle, and the square.

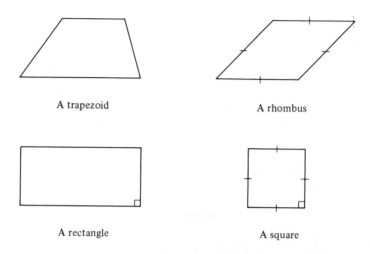

| A trapezoid | A rhombus |
| A rectangle | A square |

Figure 6-19

A quadrilateral is a *trapezoid* if at least one pair of opposite sides is parallel. There is much disagreement over the definition of a trapezoid. Sometimes a trapezoid is defined to be a quadrilateral that has one pair of opposite sides parallel and the other pair non-parallel. This requirement seems rather artificial, like demanding that an isosceles triangle have two and only two sides congruent. Our definition does not rule out the possibility that both pairs of opposite sides of a trapezoid are parallel. Thus, to us a parallelogram is a special case of a trapezoid, just as an equilateral triangle is a special case of an isosceles triangle.

A quadrilateral is a *rhombus* if it is a parallelogram all of whose sides are congruent.

A quadrilateral is a *rectangle* if it is a parallelogram all of whose angles are right angles.

A quadrilateral is a *square* if it is a rectangle all of whose sides are congruent.

We can picture the relations between these figures in the following diagram:

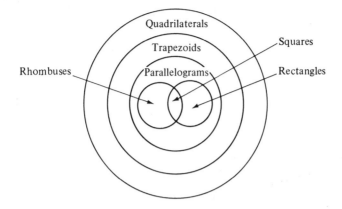

Classifying quadrilaterals

Figure 6-20

Exercise 6.5

1. Prove that either diagonal of a parallelogram divides the parallelogram into two congruent triangles.

2. Devise a "paper cutting" method of demonstrating the theorem stated in problem 1.

3. Prove Theorem 8.

4. Prove Theorem 9.

5. Prove in detail the last statement in the proof of Theorem 10.

6. Complete the proof of Theorem 11.

7. If the measure of one angle of a parallelogram is 65, find the measures of each of the other three angles.

8. Is a polygon having all of its sides congruent and all of its angles right angles necessarily a square? What additional property is needed to make this a true statement?

9. Prove that if a parallelogram contains *one* right angle, then it must be a rectangle.

10. Let *abcd* be a rhombus, *ac* a diagonal. Prove that $\angle 1 \cong \angle 2$, i.e., the diagonal bisects $\angle a$. (See the figure following.)

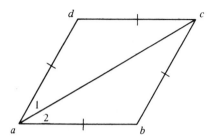

11. If the measure of one angle of a rhombus is 50, find the measures of the angles of the triangles formed by drawing

    (a) the longer diagonal.   (b) the shorter diagonal.

12. The measure of one angle of a parallelogram is twice that of a consecutive angle. Find the measure of each angle of the parallelogram.

13. State *always* if the statement is always true, *sometimes* if sometimes true, and *never* if never true.

    (a) A parallelogram is a quadrilateral.
    (b) A quadrilateral is a parallelogram.
    (c) A rhombus is a trapezoid.
    (d) A square is a rectangle.
    (e) A rhombus is a square.
    (f) A square is a rhombus.
    (g) A quadrilateral with no sides parallel is a rhombus.
    (h) A quadrilateral with exactly one pair of opposite sides congruent is a parallelogram.
    (i) The parallel sides of a trapezoid are congruent.
    (j) If a polygon has three distinct diagonals, it is a quadrilateral.

14. In Theorem 11 we proved that if a quadrilateral has opposite sides congruent, then it is a parallelogram. Do you think it is true that if a quadrilateral has opposite *angles* congruent it is a parallelogram? If so, outline a proof.

15. Draw a triangle abc. Through each vertex construct a line parallel to the opposite side. Extend these lines until they intersect in pairs. Label these points p, q, r. You now have three new triangles, $\triangle pac$, $\triangle aqb$, and $\triangle brc$. Prove that each of these triangles you have constructed is congruent to $\triangle abc$. See the figure following. (Note that this construction has produced three parallelograms. Name them.)

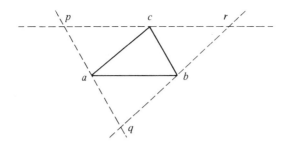

16. Draw a number of trapezoids of different sizes and shapes which are not parallelograms. In each construct a segment joining the mid-points of the two non-parallel sides. Compare the measure of this segment and the sum of the measures of the two parallel sides. Do you notice any relationship between these two numbers? State your conclusions in the form of a tentative theorem.

6. Non-Euclidean Geometry

For two thousand years Euclid's geometry was regarded as absolute truth. Based on ten axioms so self-evident that no reasoning man would deny them, and reinforced by centuries of experience and applications, Euclid's *Elements* were inviolate.

Beneath the surface, however, all was not well. A few thinkers, one of whom was Euclid himself, were disturbed by the fifth axiom, Euclid's Parallel Axiom. The parallel axiom states that given a line L and a point p not on L one and only one line may be drawn through p parallel to L. Now clearly this is an axiom that cannot be verified experimentally. We could easily draw through p two or more lines which do not intersect L on the page, no matter how large a sheet of paper we use. We must *assume* that all but one of these lines intersect L somewhere—indefinitely far away.

While the other axioms are simple, clear and intuitively reasonable, this one (in its original form—see Section 2) was lengthy, complicated and not at all obvious. Objectors felt that this statement should be proved from the other nine axioms. Its presence as an axiom was considered to be a serious flaw in an otherwise elegant and beautifully developed structure.

To resolve these doubts, many mathematicians tried to prove the parallel axiom, but all of these attempts were unsuccessful. In the nineteenth century it became apparent why this was so. Three men, Gauss, Lobatchevsky, and Bolyai, independently and within a few years of each other, discovered the geometric system which we call non-Euclidean geometry, and thereby proved that the Parallel Axiom could not be deduced from the other nine.

In trying to prove the parallel axiom, the argument went like this: Assume the other nine axioms hold but the parallel axiom does not. Then if by valid logical deductions we can arrive at a contradiction, the parallel axiom will be proved. The denial of the parallel axiom may take two forms.

(a)   Given a line L and a point p not on L, there are *no* lines through p parallel to L, and

(b)   Given a line L and a point p not on L, *at least two* lines may be drawn through p parallel to L.

Form (b) was the one chosen by Gauss, Lobatchevsky, and Bolyai. It is often called the Lobatchevskian Parallel Axiom. They assumed (b) and the other nine axioms of Euclid. From this set of axioms they deduced many strange theorems totally different from those of Euclidean geometry, but *they did not reach any contradiction of the axioms*. The conclusion they reached was revolutionary for their time: *There are geometries different from Euclid's and just as valid*. The geometry based on parallel axiom (b) is called *Lobatchevskian geometry*.

What kind of theorems can you prove in Lobatchevskian geometry? Some theorems—those based on the nine axioms of Euclid—are, of course, just the same as in Euclidean geometry. For example, the theorem, "Vertical angles are congruent," is common to both geometries. However, theorems based on the parallel axiom take a totally different form.

In Lobatchevskian geometry, we are startled to learn, the sum of the measures of the angles of a triangle is always *less than* 180. Moreover, not all triangles have the same angle sum. The larger the area of the triangle, the smaller is its angle sum! Since larger triangles have a different angle sum, there are no similar triangles—triangles that have the same shape but different size. In the new geometry if two triangles have their angles congruent, then the triangles are congruent. In other words, there is an AAA congruence theorem for triangles. In this geometry, there are no rectangles, since if three angles of a quadrilateral measure 90 the fourth must have measure less than 90.

A model for Euclidean geometry is the plane, with our intuitive notion of point and line. Let us look at a model for Lobatchevskian geometry. The surface in Figure 6-21 is called a *pseudosphere*.

"Lines" on this surface are taken to be the shortest path between points on the surface—these paths are called *geodesics*. The behavior of points and lines on this surface fits the axioms of Lobatchevskian geometry. Given a line L and a point p not on L, infinitely many lines may be drawn through p parallel to L. On this surface parallel lines are not everywhere

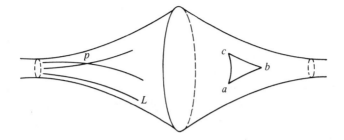

The pseudosphere, a model for Lobatchevskian geometry

Figure 6-21

equidistant—that is a theorem of Euclidean geometry. The sum of the measures of the angles of △abc is less than 180.

Not long after the invention of Lobatchevskian geometry, a mathematician named Bernhard Riemann constructed another non-Euclidean geometry, this time using axiom (a) and some (not all) of the other axioms of Euclid. His geometry was different from either the Euclidean or Lobatchevskian. In Riemannian geometry there are no parallel lines and the sum of the measures of the angles of a triangle is *greater than* 180.

A model for Riemannian geometry is the surface of a sphere. (Chapter 2, Section 5)

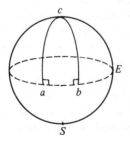

The sphere, a model for Riemannian geometry

Figure 6-22

"Lines" in this model are great circles, the geodesics of a sphere. All lines intersect and △abc contains two right angles, thus its angle sum is greater than 180. In this geometry the larger the area of the triangle, the larger its angle sum, and the only similar triangles are those which are congruent.

At this point, the student may be understandably confused. We have, not one, but three contradictory geometries. Which one is true? To be more specific, what *is* the sum of the measures of the angles of a triangle? Is it 180, less than 180, or greater than 180? To answer these questions we need to answer another more profound one. What is "truth" in mathematics? To a mathematician a theorem is "true" within a given axiomatic system if it can be deduced by valid logical arguments from the axioms. Truth in mathematics is relative—relative to the axioms which you have assumed. Thus, if we assume that the Euclidean Parallel Axiom holds, then relative to that axiom system, the theorem that the sum of the measures of the angles of a triangle is equal to 180 is "true." On the other hand, if we assume the Lobatchevskian Parallel Axiom instead of the Euclidean one, then in *this* geometry the theorem is false.

This raises an interesting question. In the physical world in which we live, which geometry applies, the Euclidean, the Lobatchevskian, or the Riemannian? Let us conduct an experiment. Suppose we construct a triangle, and as accurately as we can, measure its angles. Suppose we do this and find that the angle sum is 179.9. To verify our data, we repeat the experiment and find this time that the angle sum is 180.1. What does this mean? Was the geometry Lobatchevskian the first time and Riemannian the second? Of course not. Since any physical measurement can be only approximate, our experiment has shown only that the angle sum was somewhere in the neighborhood of 180, with an error depending on the accuracy of the instruments. In both Lobatchevskian and Riemannian geometries, the smaller the triangle the closer the angle sum to 180, so our experiment was inconclusive. We might try the experiment again with a very large triangle. Gauss carefully measured the angles of a triangle whose vertices were three widely separated mountain peaks, but his experiment was inconclusive also.

On the relatively flat portion of the earth where we find ourselves, Euclidean geometry serves us well. In the vast extent of space, however, the question may be an important one which is not yet answered. In his theory of relativity Einstein assumed that the geometry of the universe was Riemannian.

New Terms Found in This Chapter

Term	Section	Term	Section
transversal	1	distance between two	
alternate interior angles	1	parallel lines	4
corresponding angles	1	trapezoid	5
lemma	1	rhombus	5
exterior angle of a triangle	1	rectangle	5
opposite interior angles of		square	5
a triangle	1	Lobatchevskian geometry	6
regular polygon	Exercise 6.3	pseudosphere	6
parallelogram	4	geodesics	6
distance from a point to		Riemannian geometry	6
a line	4		

7: Similarity

1. Similarity and Proportion

Intuitively speaking, we consider two objects to be similar if they have the same shape, although they may not have the same size. We would consider all squares to be similar and all circles to be similar, although we would *not* consider the two rectangles in Figure 7-1 to be similar.

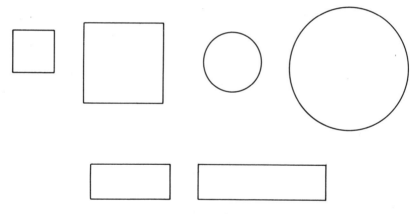

Two objects are similar if they have the same shape, although they may not have the same size.

Figure 7-1

We are using the concept of similar figures when we make scale drawings or maps. A map of a city, for instance, has many properties which are shared by the city itself. If two streets intersect in the city, the lines representing these streets intersect on the map, and at the same angle. One property which is not shared, of course, is *measure*. However, maps are usually drawn in such a way that we can find the distance between two points in the city by a simple calculation. We use what is called the *scale* of the map. For example, the scale may be 1 inch: 1 mile. This means if we measure the distance between two points on the map and find it to be $2\frac{1}{2}$ inches, then we conclude that the two corresponding points in the city are $2\frac{1}{2}$ miles apart. In this chapter we shall look more closely at the relationship of similarity between polygons, and see what properties similar figures have in common.

If we have a number of fractions which are all equivalent, such as

$$\tfrac{1}{2}; \tfrac{2}{4}; \tfrac{3}{6}; \tfrac{4}{8}; \tfrac{5}{10}$$

then we say the ratio of the numerator to the denominator is the same in each case. The set of numerators, 1, 2, 3, 4, 5, is said to be proportional to the set of denominators, 2, 4, 6, 8, 10.

Two ordered sets of positive numbers $a_1, a_2, \ldots, a_n$ and $b_1, b_2, \ldots, b_n$ are said to be *proportional* if

$$\frac{a_1}{b_1} = \frac{a_2}{b_2} = \frac{a_3}{b_3} = \cdots = \frac{a_n}{b_n} = k$$

The common ratio, k, is called the *constant of proportionality*. In the example above, the constant of proportionality is $\frac{1}{2}$.

The constant of proportionality is a generalization of the idea of the scale of a map. For example, if $a_1, a_2, \ldots, a_n$ stand for distances on a map, and $b_1, b_2, \ldots, b_n$ stand for the corresponding distances in the city, then the ratio of any two corresponding distances, say a_1 to b_1, is the scale.

If two sets are proportional, we write

$$a_1, a_2, \ldots, a_n \sim b_1, b_2, \ldots, b_n$$

For example, 1, 7, $8\frac{1}{2}$, 14 ~ 3, 21, $25\frac{1}{2}$, 42 since $\frac{1}{3} = \frac{7}{21} = 8\frac{1}{2}/25\frac{1}{2} = \frac{14}{42}$.

It is not hard to see that if $a_1, a_2, \ldots, a_n \sim b_1, b_2, \ldots, b_n$ then $b_1, b_2, \ldots, b_n \sim a_1, a_2, \ldots, a_n$, for if

$$\frac{a_1}{b_1} = \frac{a_2}{b_2} = \cdots = \frac{a_n}{b_n} = k$$

then

$$\frac{b_1}{a_1} = \frac{b_2}{a_2} = \cdots = \frac{b_n}{a_n} = \frac{1}{k}$$

For example, 1, 2, 3, 4 ~ 2, 4, 6, 8 with $k = \frac{1}{2}$, while 2, 4, 6, 8 ~ 1, 2, 3, 4 with $k = 2$.

In order to work with proportional sets, we merely express them as equations between fractions and use the laws of algebra.

Note that since $\frac{3}{4} = \frac{6}{8}$, then 3, 6 ~ 4, 8. However, it is also true that the ratio of the two numerators is the same as the ratio of the two denominators, that is $\frac{3}{6} = \frac{4}{8}$. Thus, 3, 4 ~ 6, 8. This illustrates our next theorem.

THEOREM 1   If $a, b \sim c, d$, then $a, c \sim b, d$.

Proof: If $a, b \sim c, d$, then

$$a/c = b/d \text{ by definition, and } ad = bc$$

Dividing both sides of the equation by bd (Recall that these are all positive numbers, hence none of them are zero.), we get

$$\frac{ad}{bd} = \frac{bc}{bd} \quad \text{or} \quad \frac{a}{b} = \frac{c}{d}$$

This says $a, c \sim b, d$. ∎

The proportionality $a, b \sim c, d$ is often read "a is to c as b is to d," or in view of Theorem 1, "a is to b as c is to d."

Now consider the two triangles $\triangle abc$ and $\triangle def$ whose sides have the measures given in Figure 7-2.

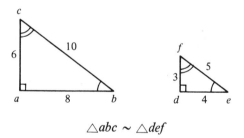

$$\triangle abc \sim \triangle def$$

Figure 7-2

Under the correspondence $abc \longleftrightarrow def$ the side $\overline{ab}$ corresponds to $\overline{de}$, $\overline{cb}$ corresponds to $\overline{fe}$ and so on. The ratios of the lengths of these corresponding sides is a constant since $\frac{6}{3} = \frac{8}{4} = \frac{10}{5} = 2$. Thus, 6, 8, 10 ~ 3, 4, 5, and we say that *corresponding sides are proportional.*

Two polygons are said to be *similar* if there is a one-to-one correspondence between their vertices such that (1) corresponding angles are congruent, and (2) corresponding sides are proportional. We will use the symbol "~" for the relation "is similar to." The two triangles in the example are similar, and we write

$$\triangle abc \sim \triangle def$$

Just as for congruences, the notation indicates the correspondence. When we write $\triangle abc \cong \triangle def$ (congruence), we mean

$$\angle a \cong \angle d \qquad\qquad \overline{ab} \cong \overline{de}$$
$$\angle b \cong \angle e \quad \text{and} \quad \overline{bc} \cong \overline{ef}$$
$$\angle c \cong \angle f \qquad\qquad \overline{ac} \cong \overline{df}$$

When we write $\triangle abc \sim \triangle def$ (similarity), we mean

$$\angle a \cong \angle d$$
$$\angle b \cong \angle e$$
$$\angle c \cong \angle f$$

and $ab, bc, ac \sim de, ef, df$; i.e., $ab/de = bc/ef = ac/df$.

Similar polygons, of course, are not necessarily congruent. However, congruent polygons are similar. If two polygons are congruent, corresponding angles are congruent and *corresponding sides are congruent as well.* If corresponding sides are congruent, then obviously they are proportional with constant of proportionality equal to one.

Note that for two polygons to be similar, *two* conditions must be satisfied. If only one condition is satisfied, this is not in general enough to guarantee that the two figures will have the same shape. For example, in the two rectangles in Figure 7-3, condition (1) is satisfied. Under any correspondence the angles are congruent since they are all right angles. However, the two figures are not similar, since no matter what correspondence we choose, the sides will not be proportional.

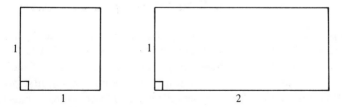

These two rectangles are not similar even though all angles are congruent.

Figure 7-3

Now consider the rhombus and the square in Figure 7-4.

Under any correspondence corresponding sides are proportional, even congruent; however, the two figures are not similar since none of the angles are congruent.

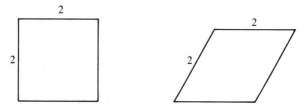

These two quadrilaterals are not similar even though all sides are congruent (hence proportional).

Figure 7-4

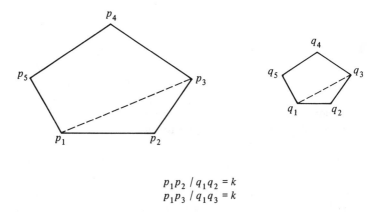

$$p_1p_2 \,/\, q_1q_2 = k$$
$$p_1p_3 \,/\, q_1q_3 = k$$

Corresponding diagonals of similar polygons are proportional.

Figure 7-5

Suppose that we have two similar polygons, $p_1p_2p_3p_4p_5$ and $q_1q_2q_3q_4q_5$ and suppose the ratio of two corresponding sides is k. (Figure 7-5)

We can think of one of these polygons as a scale drawing of the other, and it is not hard to see that given any two points on the first polygon the ratio of the length of the segment determined by these two points to the length of the segment determined by the corresponding points on the second polygon will be equal to k. Thus, corresponding diagonals of similar polygons are proportional with the same constant of proportionality as the corresponding sides. Corresponding medians of similar triangles are proportional, as are corresponding altitudes.

The *perimeter* of a polygon is defined as the sum of the measures of its sides. Suppose that we have two similar polygons whose sides measure

$$s_1, s_2, s_3, \ldots, s_n \qquad \text{and} \qquad t_1, t_2, t_3, \ldots, t_n,$$

respectively. Then

$$\frac{s_1}{t_1} = \frac{s_2}{t_2} = \cdots = \frac{s_n}{t_n} = k$$

or $$s_1 = kt_1;\; s_2 = kt_2;\; \ldots;\; s_n = kt_n$$

The perimeter of the first polygon then is

$$(s_1 + s_2 + \cdots + s_n) = kt_1 + kt_2 + \cdots + kt_n$$
$$= k(t_1 + t_2 + \cdots + t_n)$$

The ratio of the two perimeters is

$$\frac{s_1 + s_2 + \cdots + s_n}{t_1 + t_2 + \cdots + t_n} = k$$

Exercise 7.1

1. Show that the following pairs of ordered sets are proportional. Find the constant of proportionality.

   (a) 1, 3, 7 and 3, 9, 21
   (b) 2, 4, 7, 9, 15 and 8, 16, 28, 36, 60
   (c) 3, 21, 30, 33 and 1, 7, 10, 11
   (d) $\frac{1}{2}, \frac{1}{3}, \frac{1}{4}, \frac{1}{5}$ and $\frac{1}{3}, \frac{2}{9}, \frac{1}{6}, \frac{2}{15}$
   (e) $\frac{2}{5}, \frac{1}{4}, \frac{3}{8}, \frac{1}{3}$ and $\frac{3}{10}, \frac{3}{16}, \frac{9}{32}, \frac{1}{4}$
   (f) $x, 2x, 3x$ and $3y, 6y, 9y;\; x > 0;\; y > 0$
   (g) $\sqrt{2}, \sqrt{3}, \sqrt{6}$ and $2, \sqrt{6}, 2\sqrt{3}$

2. In the following, find all pairs of ordered sets that are proportional.

   (a) 1, 2, 3   (b) 9, 7, 17
   (c) 5, 7, 9   (d) 18, 14, 34
   (e) 27, 21, 51   (f) 15, 30, 45
   (g) 10, 14, 18   (h) $2\frac{1}{2}, 3\frac{1}{2}, 4\frac{1}{2}$
   (i) $\frac{1}{3}, \frac{2}{3}, 1$

3. If $2, 5, 13 \sim 1, x, y$, find x and y.

4. If $3, x, 12 \sim 1, 2, y$, find x and y.

5. The sides of a triangle measure 3, 5 and 7 inches, respectively. If the shortest side of a similar triangle measures 4 inches, find the measures of the other two sides.

6. Suppose the polygons $p_1p_2p_3p_4p_5$ and $q_1q_2q_3q_4q_5$ are similar under

the correspondence $p_1p_2p_3p_4p_5 \longleftrightarrow q_1q_2q_3q_4q_5$; $p_1p_2 = 2$; $p_2p_3 = 5$; $p_3p_4 = 1$; $p_4p_5 = 6$; $p_5p_1 = \frac{1}{2}$. If the ratio of proportionality is $\frac{3}{2}$, find the measures of all the sides of $q_1q_2q_3q_4q_5$.

7. The sides of a polygon measure 1, 3, 5, 6, and 7, respectively. Find the measures of the sides of a similar polygon if its perimeter is 33.

8. The sides of a polygon measure $\frac{1}{2}$, 1, $2\frac{1}{2}$, 3, and 4, respectively. If the longest side of a similar polygon has measure 20, find its perimeter.

9. (a) Show that the proportionality relation "$\sim$" is reflexive. That is, show that $a_1, a_2, \ldots, a_n \sim a_1, a_2, \ldots, a_n$.

   (b) Show that the proportionality relation "$\sim$" is transitive. That is, if $a_1, a_2, \ldots, a_n \sim b_1, b_2, \ldots, b_n$ and $b_1, b_2, \ldots, b_n \sim c_1, c_2, \ldots, c_n$ show that $a_1, a_2, \ldots, a_n \sim c_1, c_2, \ldots, c_n$. (*Hint:* Suppose $a_1/b_1 = j$ and $b_1/c_1 = k$. Then $a_1 = b_1 j$ and $c_1 = b_1/k$, and $a_1/c_1 = b_1 j/(b_1/k) = jk$.)

A number of years ago, mathematics students were drilled extensively in ratio and proportion and the "rules of four." Theorem 1 was one of these rules. It was called the Rule of Alternation. Prove the rules given below in problems 10–12.

10. The Rule of Composition. If $a, b \sim c, d$ then $a + b, b \sim c + d, d$. (*Hint:* Use Theorem 1. Express $a, c \sim b, d$ as an equation between fractions and then add 1 to both sides of the equation.)

11. The Rule of Division. If $a, b \sim c, d$ then $a - b, b \sim c - d, d$.

12. The Rule of Composition and Division. If $a, b \sim c, d$ then $a + b, a - b \sim c + d, c - d$.

13. Give numerical examples to illustrate each of the rules given in Problems 10–12.

14. Using Theorem 1 and the rules of problems 10–12 complete the following statements:

    (a) If $a/b = \frac{2}{3}$, then $a/2 =$ _____.
    (b) If $x/2 = y/3$, then $(x + 2)/2 =$ _____.
    (c) If $(2 + 3)/3 = (x + y)/y$, then $\frac{2}{3} =$ _____.
    (d) If $x/y = \frac{1}{3}$, then $(x + y)/(x - y) =$ _____.
    (e) If $(a + 2)/(a - 2) = (b + 4)/(b - 4)$, then $a/2 =$ _____.
    (f) If $\frac{3}{2} = x/y$, then $(3 - 2)/2 =$ _____.

2. Similarity of Triangles

We have seen that in general two polygons are similar if and only if *both* of the following conditions hold:

(1) Corresponding angles are congruent.

(2) Corresponding sides are proportional.

For triangles the situation is simpler. *Either one of these conditions is sufficient alone to guarantee similarity.* Intuitively you may feel that the shape of a triangle is determined by its angles alone, and this is true.

THEOREM 2 *The AAA Similarity Theorem.* If corresponding angles of two triangles are congruent, then the triangles are similar.

Thus, if we know that corresponding angles of two triangles are congruent, we can conclude that corresponding sides are proportional.

By the Angle Sum Theorem (Chapter 6) if we know the measures of two angles of a triangle we know the measure of the third.

THEOREM 3 *The AA Similarity Theorem.* If two pairs of corresponding angles of two triangles are congruent, then the triangles are similar.

Consider, for example, the triangles *abc* and *def* in Figure 7-6 where $\angle a \cong \angle d$; $\angle b \cong \angle e$.

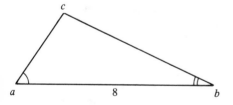
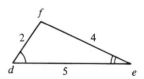

If $\angle a \cong \angle d$ and $\angle b \cong \angle e$, then $\triangle abc \sim \triangle def$.

Figure 7-6

The AA Similarity Theorem tells us that $\triangle abc \sim \triangle def$. If they are similar, then corresponding sides are proportional,

$$ab, bc, ca \sim de, ef, fd$$

Thus, if we know the lengths of the sides of the smaller triangle and one of the sides of the larger triangle, we can find the measures of the other two sides of $\triangle abc$. Suppose, for example, $de = 5$, $ef = 4$, $fd = 2$, and $ab = 8$. Then

$$5, 4, 2 \sim 8, bc, ca$$

or
$$\frac{5}{8} = \frac{4}{bc} = \frac{2}{ca}$$

Solving the first equation, we get

$$5bc = 32$$
$$bc = \tfrac{32}{5} = 6.4$$

In the same way

$$5ca = 16$$
$$ca = \tfrac{16}{5} = 3.2$$

The following theorem says that for two triangles, having proportional sides alone is sufficient for similarity.

THEOREM 4   *The SSS Similarity Theorem.* If corresponding sides of two triangles are proportional, then the triangles are similar.

We also have an analogue of the SAS congruence axiom.

THEOREM 5   *The SAS Similarity Theorem.* If there is a correspondence between two triangles such that two pairs of corresponding sides are proportional and the included angles are congruent, then the triangles are similar.

The theorems in this section have many applications. Suppose that you wish to find the height of a tree. Your height is 6 ft., and at a certain hour on a sunny day you cast a 4 ft. shadow. At the same time of day, the tree's shadow is 10 ft. long. (Figure 7-7)

If we assume that the triangles formed, $\triangle abc$ and $\triangle def$ are right triangles, then since the rays of the sun are parallel, $\overline{ac} \; /\!/ \; \overline{df}$, and therefore, $\angle a \cong \angle d$. (Why?) Thus by the AA Similarity Theorem, $\triangle abc \sim \triangle def$ and corresponding sides are proportional. We are interested in particular in the horizontal and vertical sides of the triangles. We have $4, 6 \sim 10, fe$, hence $\frac{4}{10} = 6/fe$.

$$4fe = 60$$
$$fe = 15$$

We conclude that the tree is 15 ft. tall.

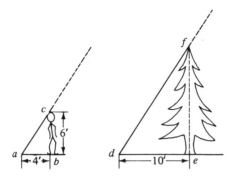

An application of the AA Similarity Theorem

Figure 7-7

An *altitude* of a triangle is a perpendicular line segment from a vertex to the line containing the opposite side. Every triangle has three altitudes. (Figure 7-8)

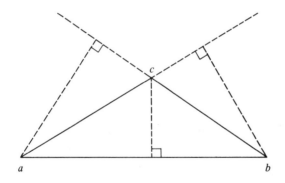

The three altitudes of a triangle

Figure 7-8

An altitude may lie in the interior or in the exterior of a triangle; however, in a right triangle, the altitude from the vertex of the right angle will always lie in the interior.

In a right triangle, the side opposite the right angle is called the *hypotenuse* and the other two sides are referred to as *legs*.

THEOREM 6 The altitude from the vertex of the right angle of a right triangle divides the triangle into two right triangles, each of which is similar to the original triangle. (Figure 7-9)

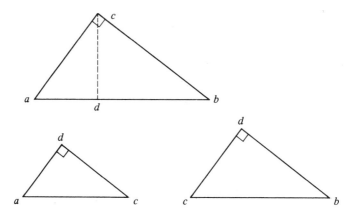

The altitude from the vertex of the right angle of a right triangle divides the triangle into two right triangles, each of which is similar to the original triangle.

Figure 7-9

Let *abc* be a right triangle with m $\angle c = 90$, and let $\overline{cd}$ be the altitude from *c*. (Figure 7-9) Then

$$\triangle abc \sim \triangle acd$$

and $\qquad\qquad \triangle abc \sim \triangle cbd$

Proof: Angle *a* is congruent to itself, and $\angle c \cong \angle adc$ since they are both right angles. Then by the AA Similarity Theorem $\triangle abc \sim \triangle acd$. In the same way, we can show that $\triangle abc \sim \triangle cbd$. ∎

It is not hard to keep these correspondences straight, since there is only one way they can be set up. For example, in finding the correspondence between $\triangle abc$ and $\triangle adc$, $\angle a$ must correspond to $\angle a$, and $\angle c$, which is a right angle, must correspond to the right angle whose vertex is at *d*. Then *b* must correspond to *c* as this is the only place it can go. Thus we have:

$$a \longleftrightarrow a \qquad b \longleftrightarrow c \qquad c \longleftrightarrow d$$

or $\qquad\qquad \triangle abc \sim \triangle acd$

Since each of the small right triangles is similar to the large one, the two small triangles must be similar to each other. (This is true because similarity between triangles is an equivalence relation. (See Exercise 7.2, problem 2.) Thus,

$$\triangle abc \sim \triangle acd$$

and $\qquad\qquad\qquad \triangle abc \sim \triangle cbd$

implies that $\qquad\qquad \triangle acd \sim \triangle cbd$

    If we cut $\triangle abc$ apart and separate these two triangles, the similarity is easier to see. (Figure 7-9)

Exercise 7.2

1. Given $\triangle abc \sim \triangle def$. Prove that if one pair of corresponding sides is congruent, then the two triangles are congruent.

2. Show that the relation "is similar to" on the set of triangles is an equivalence relation.

3. Let abc be any triangle, with $\overline{de} \parallel \overline{ab}$. Prove that $\triangle abc \sim \triangle dec$.

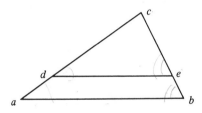

4. Can two triangles be similar if the first contains an angle of measure 70 and the second contains an angle of measure 115? Explain.

5. Are triangles abc and def similar if m $\angle a = 62$, m $\angle b = 54$, m $\angle d = 54$, and m $\angle e = 64$? If so, what is the correspondence?

6. Is it possible for two triangles to be similar if the first contains angles measuring 45 and 72 and the second contains angles measuring 72 and 85? Explain.

7. If the lengths of the sides of $\triangle abc$ and $\triangle def$ are $ab = 5$, $ac = 3$, $bc = 7$, and $de = 9$, $df = 15$, $ef = 21$, are the triangles similar? If so, give the correspondence.

8. If one angle of an isosceles triangle is congruent to an angle of a second isosceles triangle, are the two triangles necessarily similar? Explain.

9. Prove that if a pair of acute angles of two right triangles are congruent, then the two triangles are similar.

10. Prove that any two equilateral triangles are similar.

11. Prove that any two isosceles right triangles are similar.

12. A tower $\overline{ac}$ casts a shadow 100 feet long. A 9 foot pole is placed upright at d so that the tip of its shadow coincides with the tip of the shadow of the tower. The distance $\overline{db}$ is measured and found to be 12 feet. How tall is the tower?

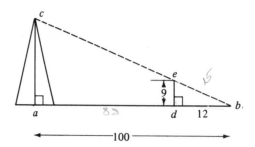

13. A flagpole $\overline{pq}$ casts a shadow $\overline{ps}$. A man 6 feet tall stands 18 feet from the pole, and his shadow $\overline{rs}$ is 7 feet long. How tall is the flagpole?

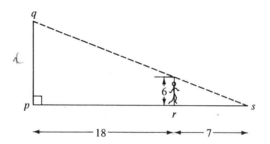

14. Let L_1, L_2, and L_3 be parallel lines, and let the distance from L_1 to L_2 be two units and the distance from L_2 to L_3 be one unit. Let T be a transversal cutting L_1, L_2, and L_3 in points p_1, p_2, and p_3. Prove that segment $\overline{p_1 p_2}$ is twice as long as $\overline{p_2 p_3}$.

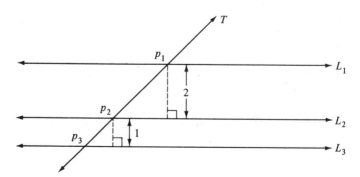

15. Prove that if in $\triangle abc$, d is the midpoint of $\overline{ac}$ and e is the midpoint of $\overline{cb}$, then $\overline{de}$ // $\overline{ab}$ and $de = \frac{1}{2}ab$. (*Hint:* Use the SAS Similarity Theorem to show that $\triangle abc \sim \triangle dec$.)

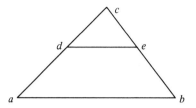

3. The Pythagorean Theorem

The Pythagorean Theorem must surely be the most famous theorem in geometry. Although special cases of the theorem were known long before the time of Pythagoras, he probably was the first to give a general proof of the theorem. Over 2,000 years before the birth of Christ, Egyptian surveyors used an application of this theorem to construct a right angle. This simple device was a rope with two knots in it. The knots divided the rope into three sections whose lengths were proportional to the numbers 3, 4, 5. When the rope was staked out to form a triangle with two vertices at the knots, the surveyor had constructed a right angle at knot a.

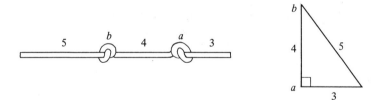

Constructing a right angle using a rope with two knots in it.

Figure 7-10

Hundreds of proofs of the Pythagorean Theorem have been given. In the following proof we will use the properties of similar triangles developed in this chapter. To simplify our notation we will use capital letters A, B, C for the vertices of our triangle and small letters, a, b, c for the lengths of the sides opposite these vertices. Thus, a is the length of the side opposite vertex A, and so on.

THEOREM 7   *The Pythagorean Theorem.* In any right triangle the square of the length of the hypotenuse is equal to the sum of the squares of the lengths of the other two sides.

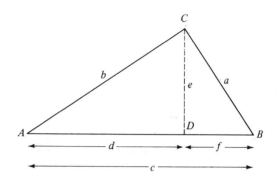

The Pythagorean Theorem. If $\angle C$ is a right angle, then $a^2 + b^2 = c^2$.

Figure 7-11

Proof: Given $\triangle ABC$, with m $\angle C = 90$, we want to show that $c^2 = a^2 + b^2$.

Let $\overline{CD}$ be the altitude from vertex C. Then by Theorem 5, $\triangle ACD \sim \triangle ABC \sim \triangle CBD$.

Thus, corresponding sides are proportional, and

$$e, d, b \sim a, b, c \sim f, e, a$$

Thus,
$$\frac{e}{a} = \frac{d}{b} = \frac{b}{c}$$

Using the last equality we have

$$\frac{d}{b} = \frac{b}{c} \quad \text{or} \quad d = \frac{b^2}{c}$$

Moreover,
$$\frac{a}{f} = \frac{b}{e} = \frac{c}{a}$$

Using the equality of the first and last terms, we have

$$\frac{a}{f} = \frac{c}{a} \quad \text{or} \quad \frac{1}{f} = \frac{c}{a^2}, \quad \text{hence} \quad f = \frac{a^2}{c}$$

Since $d + f = c$,

$$\frac{b^2}{c} + \frac{a^2}{c} = c$$

and $$b^2 + a^2 = c^2 \quad \blacksquare$$

The converse of the Pythagorean Theorem is also true.

THEOREM 8 *Converse of the Pythagorean Theorem.* In a triangle whose sides have lengths a, b, and c, if $a^2 + b^2 = c^2$, then the triangle is a right triangle with right angle opposite the side of length c.

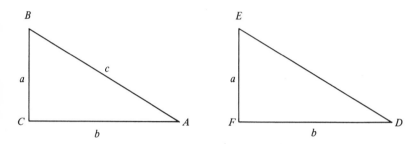

The converse of the Pythagorean Theorem. If $a^2 + b^2 = c^2$ then $\angle C$ is a right angle.

Figure 7-12

Proof: Let $\triangle ABC$ have sides of lengths a, b, and c, such that $a^2 + b^2 = c^2$. Construct a right angle at F and line segments $\overline{FD}$ and $\overline{FE}$ of lengths b and a, respectively.

Then, since $\triangle DEF$ is a right triangle, by the Pythagorean Theorem, $a^2 + b^2 = (ED)^2$, or $ED = \sqrt{a^2 + b^2}$.

But $c = \sqrt{a^2 + b^2}$ also, hence $ED = c$.

By the SSS Axiom then, $\triangle ABC \cong \triangle DEF$ and m $\angle C =$ m $\angle F = 90$. $\blacksquare$

It was actually this converse that the Egyptian surveyors used, for they constructed a triangle with sides measuring 3, 4, and 5 units, respectively. Since $3^2 + 4^2 = 5^2$ this triangle was a right triangle with right angle opposite the side having length 5.

Example:

If the legs of a right triangle are 5 and 12, respectively, find the length of the hypotenuse. Find the length of the altitude from the right angle. (Figure 7-13)

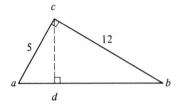

Using Theorem 6 and the Pythagorean Theorem to find the length of the altitude $\overline{cd}$

Figure 7-13

By the Pythagorean Theorem,

$$(ab)^2 = 12^2 + 5^2 = 169$$

and $$ab = 13$$

Since by Theorem 5, $\triangle abc \sim \triangle cbd$, then $cd, db, 12 \sim 5, 12, 13$

and $$\frac{cd}{5} = \frac{12}{13}$$

$$cd = \frac{(5 \cdot 12)}{13} = 4\frac{8}{13}$$

Exercise 7.3

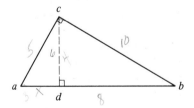

In the figure above, $\triangle abc$ is a right triangle; m $\angle c = 90$; $\overline{cd}$ is the altitude from c. (*Hint:* Use Theorem 6 and the Pythagorean Theorem.)

1. If $ad = 3$, $cd = 4$, find ac, cb, and db.

2. If $cb = 10$, $cd = 6$, find db, ac, and ad.

3. If $ad = 2$, $db = 8$, find ac, cb, and cd.

4. If $ac = 9$, $cb = 12$, find ab, ad, and cd.

5. Suppose that you have a large piece of wood and you want to know if its corners form right angles. You don't have a protractor or a car-

penter's square, but you do have a tape measure and a pencil. Explain how you might determine whether or not the angles are right angles.

6. (The Hypotenuse-Leg Theorem) Prove that if the hypotenuse and a leg of one right triangle are congruent to the hypotenuse and a leg of a second right triangle, then the triangles are congruent.

7. In an isosceles right triangle, the length of the hypotenuse is 4. Find the lengths of the legs.

8. Let $\triangle abc$ be an equilateral triangle with sides having measure s. What is the length of the altitude h in terms of s?

9. Find x and y in the following figure.

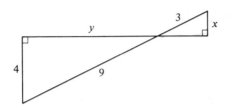

10. A set of three positive numbers, such as $3, 4, 5$, which can be the measures of the sides of a right triangle is called a *Pythagorean triple*. If (a, b, c) is a Pythagorean triple $(a^2 + b^2 = c^2)$, show that $(2a, 2b, 2c)$ is also a Pythagorean triple. For any positive integer n, show that (na, nb, nc) is a Pythagorean triple. Use this to construct other triples starting with $(3, 4, 5)$.

11. The Greeks knew that a Pythagorean triple (a, b, c) could be found by assigning values to u and v, where

    (1) u and v have no factors in common,

    (2)   one is even, the other odd,

and

    (3) $u > v$

and using the formulas

$$a = 2uv; \quad b = u^2 - v^2; \quad c = u^2 + v^2$$

For example, if $u = 2$, $v = 1$, then $a = 4$, $b = 3$, $c = 5$. Use these formulas to find other Pythagorean triples.

4.   Trigonometry

The word trigonometry comes from two Greek words meaning the measurement of triangles. In this section we will look at some of the ideas involved in the study of trigonometry, and how they can be used to solve some rather interesting problems.

Two right triangles are similar if an acute angle of one is congruent to an acute angle of the other (Exercise 7.2, problem 9). Thus in the figure below the right triangles *abc* and *ade* are similar since they share angle *a*.

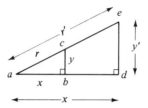

Since $\triangle abc \sim \triangle ade$, $x, y, r \sim x', y', r'$ and $y/x = y'/x'$; $y/r = y'/r'$; $x/r = x'/r'$.

Figure 7-14

If we let x, y, and r stand for the lengths of the sides of $\triangle abc$ and x', y', and r' the lengths of the corresponding sides of $\triangle ade$, then $x, y, r \sim x', y', r'$. Therefore, $x/x' = y/y'$, and $y/x = y'/x'$.

The ratio of the length of the leg opposite $\angle a$ to the length of the leg adjacent to $\angle a$ is the same for any right triangle having an angle congruent to $\angle a$.

This ratio clearly depends only on the measure of $\angle a$, not on the lengths of the sides of the triangle. We call this ratio the *tangent of angle a*, and write this "tan a." Thus,

$$\tan a = \frac{y}{x} = \frac{\text{length of the leg opposite } \angle a}{\text{length of the leg adjacent to } \angle a}$$

In the same way we can show that $y/r = y'/r'$ and $x/r = x'/r'$. These ratios depend only on the measure of $\angle a$ and are called the *sine of $\angle a$*, written "sin a," and the *cosine of $\angle a$*, written "cos a," respectively.

$$\sin a = \frac{y}{r} = \frac{\text{length of the leg opposite } \angle a}{\text{length of the hypotenuse}}$$

$$\cos a = \frac{x}{r} = \frac{\text{length of the leg adjacent to } \angle a}{\text{length of the hypotenuse}}$$

If $\angle a$ has measure n, we will also write sin $n°$, cos $n°$, or tan $n°$ for the appropriate ratios.*

---

*There are three other trigonometric ratios: cosecant $a = \csc a = r/y$; secant $a = \sec a = r/x$; and cotangent $a = \cot a = x/y$.

These ratios can be computed approximately by constructing a right triangle with one of its angles having a given measure, say 1. Then by carefully measuring the sides of the triangle, the various ratios can be found for an angle of measure 1.

Actually these ratios have been computed with great accuracy by using more advanced mathematical methods, and are available to us in tables such as the one on p. 162.

Thus for example, from the table we see that sin 20° = .342, tan 31° = .601, and cos 16° = .961. These are, of course, approximations since these trigonometric ratios may not be rational numbers.

For some special cases these ratios can be found exactly. For example, consider the case of the right triangle, one of whose angles has measure 45. Then the other acute angle has measure 45 also and the triangle must be isosceles.

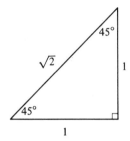

$$\sin 45° = \cos 45° = 1/\sqrt{2} \doteq .707$$
$$\tan 45° = 1/1 = 1$$

Figure 7-15

If we take the length of each leg to be one unit, then by the Pythagorean Theorem, the length of the hypotenuse must be $\sqrt{1^2 + 1^2} = \sqrt{2}$.

Then we have

$$\sin 45° = \frac{1}{\sqrt{2}} \doteq .707$$

$$\cos 45° = \frac{1}{\sqrt{2}} \doteq .707$$

$$\tan 45° = \frac{1}{1} = 1$$

(The symbol "$\doteq$" is read "is approximately equal to.")

Now consider the case of the right triangle having one acute angle of measure 60, the other of measure 30.

Table 7-1

Trigonometric Ratios

Angle	Sine	Cosine	Tangent
1°	.017	.999	.017
2°	.035	.999	.035
3°	.052	.999	.052
4°	.070	.998	.070
5°	.087	.996	.087
6°	.105	.995	.105
7°	.122	.993	.123
8°	.139	.990	.141
9°	.156	.988	.158
10°	.174	.985	.176
11°	.191	.982	.194
12°	.208	.978	.213
13°	.225	.974	.231
14°	.242	.970	.249
15°	.259	.966	.268
16°	.276	.961	.287
17°	.292	.956	.306
18°	.309	.951	.325
19°	.326	.946	.344
20°	.342	.940	.364
21°	.358	.934	.384
22°	.375	.927	.404
23°	.391	.920	.424
24°	.407	.914	.445
25°	.423	.906	.466
26°	.438	.899	.488
27°	.454	.891	.510
28°	.469	.883	.532
29°	.485	.875	.554
30°	.500	.866	.577
31°	.515	.857	.601
32°	.530	.848	.625
33°	.545	.839	.649
34°	.559	.829	.675
35°	.574	.819	.700
36°	.588	.809	.727
37°	.602	.799	.754
38°	.616	.788	.781
39°	.629	.777	.810
40°	.643	.766	.839
41°	.656	.755	.869
42°	.669	.743	.900
43°	.682	.731	.933
44°	.695	.719	.966
45°	.707	.707	1.000

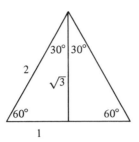

$$\sin 60° = \cos 30° = \sqrt{3}/2 \doteq .866$$
$$\cos 60° = \sin 30° = 1/2 = .5$$
$$\tan 60° = \sqrt{3}/1 \doteq 1.73$$
$$\tan 30° = 1/\sqrt{3} \doteq .577$$

Figure 7-16

If we draw two such triangles as in Figure 7-16, the larger triangle formed will be equilateral, since each of its angles will have measure 60. If we take each side to have length two units, then the shorter leg of the 30°–60° right triangle will have length one unit, and by the Pythagorean Theorem, the length of the longer leg will be $\sqrt{3}$, since

$$(\sqrt{3})^2 + 1^2 = 2^2$$

Thus we have

$$\sin 60° = \frac{\sqrt{3}}{2} \doteq .866$$

$$\cos 60° = \frac{1}{2} = .5$$

$$\tan 60° = \frac{\sqrt{3}}{1} \doteq 1.73$$

and

$$\sin 30° = \frac{1}{2} = .5$$

$$\cos 30° = \frac{\sqrt{3}}{2} \doteq .866$$

$$\tan 30° = \frac{1}{\sqrt{3}} \doteq .577$$

Notice that sin 60° = cos 30° and cos 60° = sin 30°.

In any right triangle abc with m $\angle c = 90$ (Figure 7-17), since the leg opposite $\angle a$ is adjacent to $\angle b$, and the leg adjacent to $\angle a$ is opposite to

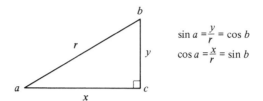

$$\sin a = \frac{y}{r} = \cos b$$

$$\cos a = \frac{x}{r} = \sin b$$

Figure 7-17

$\angle b$, the sine of $\angle a$ will be equal to the cosine of $\angle b$, and the cosine of $\angle a$ will be the sine of $\angle b$. Since in any right triangle the acute angles are complementary, we have the following rule:

The sine of an acute angle is equal to the cosine of its complement and its cosine is equal to the sine of its complement.

This means that to find the sine or cosine of any acute angle, we only need tables of these ratios for angles up to 45°. For example, to find sin 70°, we can look up cos 20°; to find cos 85°, we look for sin 5°.

Exercise 7.4

1. In the figure find sin a; cos a; tan a; sin b; cos b; tan b.

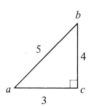

2. In the figure find sin a; cos a; tan a; sin b; cos b; tan b.

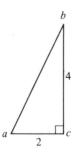

3. Let $\triangle abc$ be an isosceles right triangle with leg of length n and m $\angle c = 90$.
    What is the measure of angle a? Find the length of the hypotenuse, and compute sin a, cos a, tan a.

4. If you know the tangent of $\angle a$, how could you use this information to compute the tangent of the complement of $\angle a$? Use the following to find the tangent of the complement of $\angle a$.

   (a) $\tan a = \frac{3}{4}$ (b) $\tan a = \frac{1}{2}$

   (c) $\tan a = .25$ (d) $\tan a = x/y$

5. If m $\angle a$ is very small, then sin a is a very small number. Why? If m $\angle a$ gets larger and larger, but is still less than 90°, what happens to sin a? Does it get larger or smaller? Why? Can sin a ever be larger than 1? Why?

6. If m $\angle a$ is very small, then cos a is very close to 1. Why? If m $\angle a$ gets larger and larger, but is still less than 90°, what happens to cos a? Does it get larger or smaller? Why? Can cos a ever be larger than 1? Why?

7. If m $\angle a$ is very small, is tan a large or small? As m $\angle a$ gets larger and larger, but is still less than 90°, what happens to tan a? Explain. Can tan a be larger than 1? Why?

8. Use the table to find the measure of $\angle a$ if:

   (a) $\sin a = .438$ (b) $\cos a = .906$

   (c) $\tan a = .727$ (d) $\sin a = .940$

   (e) $\cos a = .259$

   (Remember that sin $a =$ cos $(90 - a)$.)

9. In the isosceles triangle abc, find sin a, cos a, tan a. (*Hint:* Draw an altitude from c to the opposite side to get a right triangle.)

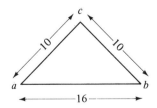

10. In an isosceles triangle abc, $ac = 10$ and m $\angle a = 70$. Find the length of the altitude to the base $\overline{ab}$ and the length of the base.

11. Use a protractor to contruct a large right triangle one of whose acute angles has measure 15. Measure all sides carefully and compute the ratios sin 15°, cos 15°, tan 15°. Compare your figures to those in the table.

5.   Indirect Measurement

It is often inconvenient or even impossible to measure distances directly. For example, in measuring the height of a mountain or the distance from the earth to the moon, a tape measure is clearly inadequate! Such distances are measured *indirectly*. To do this, we use the principles of trigonometry.

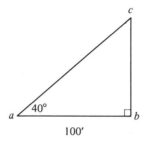

Measuring the height of a building $\overline{bc}$ indirectly
$\tan 40° = bc/100$

Figure 7-18

For example, to measure the height of a building bc (Figure 7-18), if we know that $ab = 100$ ft. and m $\angle a = 40$, then since $bc/ab = \tan a$, we can find tan 40° in our table, and use this equation to find the height of the building. Thus,

$$\tan 40° = .839 = \frac{bc}{100}$$

or $bc = 83.9$ ft.

The angle at a is called the *angle of elevation* of the top of the building, and is measured by an instrument called a *sextant*. A simple sextant can be made with a protractor and a ruler. (Figure 7-19)

The protractor is taped to the ruler as illustrated. A small weight is tied to a thread which is fastened to the center of the base of the protractor.

To measure the angle of elevation, sight along the base of the protractor. The angle of elevation is the difference between the reading at the point where the thread crosses the protractor and 90°. For example, if the thread

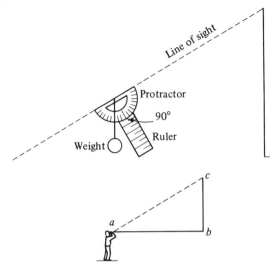

Making a simple sextant with a protractor and a ruler

Figure 7-19

crosses the protractor at 75°, the angle of elevation is 90° − 75° = 15°. The vertex of the angle you are measuring is, of course, at your eye level.

The father of trigonometry is the great astronomer-mathematician Hipparchus who lived two centuries before Christ. He discovered the facts we have discussed about the ratios associated with right triangles and computed the first trigonometry tables to five decimal places. Hipparchus was an astronomer and he used his discoveries to compute the radius of the earth, the distance from the earth to the moon and many other astronomical measurements.

Let us see how Hipparchus might have solved one of these problems using trigonometry. To compute the radius of the earth, we might first climb a mountain whose height we know to be, say 3 miles. (Figure 7-20) From the top of the mountain we sight an object on the horizon at a which is at sea level. The triangle aob will be a right triangle. We measure the angle at b and find it to be 87°46′. Since $\triangle aob$ is a right triangle,

$$\sin 87°46' = \frac{r}{r+3}$$

From a trigonometry table we find that $\sin 87°46' = .99924$ to five decimal places, so we have the equation

$$.99924 = \frac{r}{r+3}$$

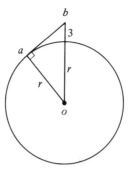

Computing the radius of the earth. (*Not* to scale)

Figure 7-20

Solving this equation for *r*, we get the value 3,944 miles for the radius of the earth.

Exercise 7.5

1.

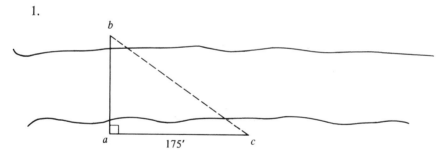

To find the width $\overline{ab}$ of a river a line segment $\overline{ac}$ 175 feet long is laid off perpendicular to $\overline{ab}$. The measure of $\angle c$ is found to be 35°. Find the width of the river.

2.

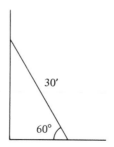

A ladder 30 feet long leans against a wall making an angle of 60° with the horizontal. How far above ground is the top of the ladder? How far from the wall is the bottom of the ladder?

3. The angle of elevation of the top of a tree is found to be 32° at a point 40 feet from the base of the tree. How tall is the tree?

4. An antenna is to be fastened by a guy wire from the top of the antenna to a post in the ground 20 feet from the base of the antenna. If the angle of elevation of the top of the antenna is 45° at the post, find the length of the guy wire.

5. A building on the bank of a river rises 175 feet above the water. If the angle of elevation of the top of the building from a point on the opposite bank of the river is 21°, find the width of the river.

New Terms Found in This Chapter

Term	*Section*	*Term*	*Section*
proportional	1	trigonometry	4
constant of proportionality	1	tangent of an angle	4
similar polygons	1	sine of an angle	4
perimeter	1	cosine of an angle	4
altitude	2	cosecant of an angle	4
hypotenuse of a right		secant of an angle	4
triangle	2	cotangent of an angle	4
legs of a right triangle	2	angle of elevation	5
Pythagorean Triple	Exercise 7.3		

8: Area

1. Introduction

In everyday English the word *area* is used in two different ways. An area can mean a region, as "Meet us in the picnic area," or more technically, an area can be a number measuring the size of a region. In mathematics the word is used in this last sense.

Intuitively, we can think of area as a measure of how large some region is. We say a house has floor area 2500 square feet, or a park is 100 square miles in area. Area is a measure in the plane—in two dimensions, just as length is a measure on a line in one dimension. To measure lengths or angles, we use an instrument with a scale—a ruler or a protractor. Area, however, is measured indirectly by using formulas. In this chapter we will develop some of these formulas.

As long ago as 2,000 B.C. the Babylonians knew some of these formulas. There is evidence that they knew the general rule for finding the area of the rectangle, the right triangle and the isosceles triangle, and a trapezoid containing one right angle.

2. Area of a Rectangle

A *polygonal region* is the union of a polygon and its interior. (Chapter 2, Section 8) Every polygonal region has associated with it a positive real number we call its area. It is common to speak of "the area of a rectangle," or "the area of a triangle," when we really mean "the area of a rectangular region" or "the area of a triangular region." We will, since this is common usage, use the phrase "area of a polygon" as an abbreviation for "area of a polygonal region."

Area has the following properties:

1. The area of a polygonal region is a positive real number.

2. If two polygonal regions are congruent, then they have the same area.

3. If two polygonal regions, A and B, intersect only in (1) vertices, or (2) sides, or (3) do not intersect at all, then the area of $A \cup B$ is the sum of the area of A and the area of B. (Figure 8-1)

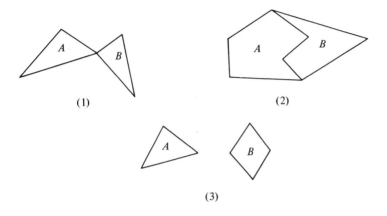

(1)   (2)

(3)

Area $(A \cup B) =$ Area $(A) +$ Area (B)

Figure 8-1

This is called the *additive property* of area. If two polygonal regions overlap, then clearly this additive property does not hold. In Figure 8-2 the shaded portion belongs to both regions and

Area $(A \cup B) <$ Area $(A) +$ Area (B)

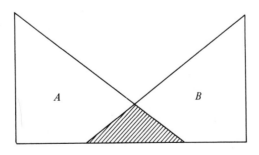

Area $(A \cup B) <$ Area $(A) +$ Area (B)

Figure 8-2

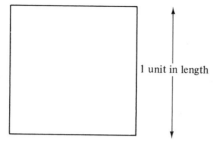

1 unit in length

1 square unit

Figure 8-3

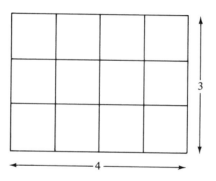

Area = 12 square units

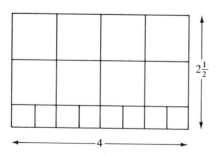

Area = 10 square units

Figure 8-4

Units of area include the square inch, the square foot, the square yard, and so on. We can think of a square unit of area as the area of a square each of whose sides has length one unit.

Intuitively then, the area of a rectangle would be the number of these square units that could be fitted snugly into the rectangle with no over-lapping. If a rectangle has adjacent sides which measure 3 and 4 units, respectively, then we can fit in 3 rows, each having 4 squares, or 12 square units altogether. (Figure 8-4)

If one of the sides of the rectangle measures, say $2\frac{1}{2}$ units, then we can cut our unit square into four smaller squares, each having area one-fourth of a square unit, and fit these smaller squares into the rectangle. (Figure 8-4)

We have 8 of the large squares, or 8 square units, plus 8 of the smaller squares, which can be reassembled to give 2 square units, or 10 square units altogether.

The only trouble with thinking of area in this way is that it doesn't work for all rectangles. For example, suppose that we have a square, each of whose sides has length $\sqrt{2}$. We could easily construct such a square by cutting apart two unit squares along their diagonals and fitting them together. (See Figure 8-5) By the Pythagorean Theorem, the length of this diagonal is $\sqrt{1^2 + 1^2} = \sqrt{2}$.

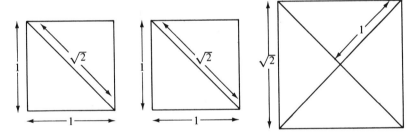

The area of the large square must be two square units, but small squares cut from these unit squares cannot be fitted snugly into the large square.

Figure 8-5

Since we constructed the largest square from the two smaller squares each having area 1 square unit, it is obvious that the large square has area 2 square units. However, there is no way in which small unit squares can be fitted into this square. Even if we cut the unit square into as many smaller squares as we please, say n^2 of them, each having a side measuring $1/n$, they cannot be fitted exactly into the large square. The reason for this is that $\sqrt{2}$ is an irrational number—it cannot be expressed as the ratio of two integers. If we could divide a side of length $\sqrt{2}$ into p pieces each having length $1/n$ (where p and n are integers), then we would have $\sqrt{2} = p/n$, which is impossible.

For this reason, although it is all right to think of the area of a rectangle *intuitively* as the number of unit squares that can be packed into it (and for rectangles with sides having rational numbers for their measures this can be done), we had better not *define* the area of a rectangle in this way. Instead we will define the area of a rectangle by a formula.

The *area of a rectangle* is the product of the lengths of two adjacent sides. One of these sides is usually called the *base* and the other the *altitude* corresponding to that base. We have then the formula

$$\text{Area of a rectangle} = b \cdot h$$

where b is the measure of the base and h the measure of the altitude corresponding to that base. (These numbers are also often called the length and the width of the rectangle.)

The terms "base" and "altitude" are commonly used in two ways. They may refer to certain line segments, or they may mean the *lengths* of these line segments. Thus we may say "The area of a rectangle is the product of its base and its altitude." When we say this we are of course thinking of the base and altitude as numbers. This kind of sloppiness is common in a branch of mathematics as old as geometry. Recall that the

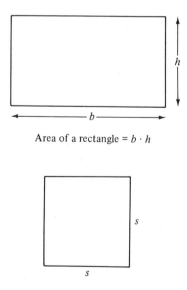

Area of a rectangle = $b \cdot h$

Area of a square = s^2

Figure 8-6

"radius of a circle" can mean a line segment or a number. There is really no harm done as long as we realize that these terms have dual meanings.

A square is a special case of the rectangle, so the area of a square is given by the same formula. Since in a square the altitude and the base have the same measure, we call this length s, and the formula becomes:

$$\text{Area of a square} = s \cdot s = s^2$$

This formula fits in with our observations, since the square with side $\sqrt{2}$ has area $(\sqrt{2})(\sqrt{2}) = 2$ square units.

Exercise 8.2

1. Complete the following table for rectangles:

Base	Height	Area
10′	7′	
$2\frac{1}{2}''$	$5\frac{1}{4}''$	
12′		144 sq. ft.
	15″	200 sq. in.

2. What is the effect on the area of a square if a side is doubled in length? If a side is tripled? If a side is cut in half? Make sketches to illustrate your answers.

3. What is the effect on the area of a rectangle if its base is doubled and its height left unchanged? If its base and height are both doubled? If its base is doubled and its height cut in half? Make sketches to illustrate your answers.

Use the additive property of area to find the area of each of the following regions:

4.

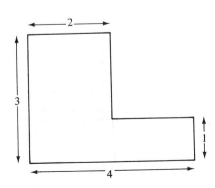

5.

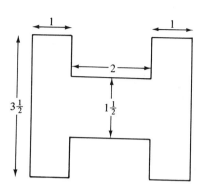

6.

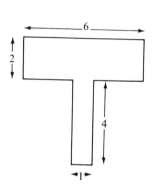

7.

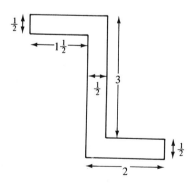

8.

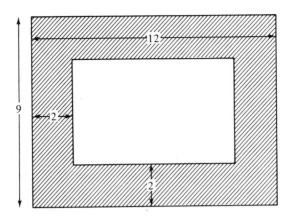

Find the area of the shaded region.

9. Consider the square whose side measures a and whose area is a^2. Explain how you might cut this square into four pieces which can be reassembled to make two squares, each having area $\frac{1}{2}a^2$. What is the measure of a side of these smaller squares?

10. The second property of area says that if two polygons are congruent then they enclose equal areas. What is the converse of this statement? Is the converse true or false? Explain.

11. A room has the dimensions:

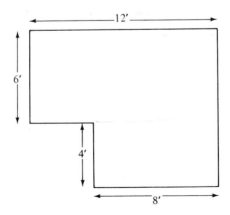

You wish to cover the floor with square tiles which measure 9″ on a side. How many tiles will you need?

12.

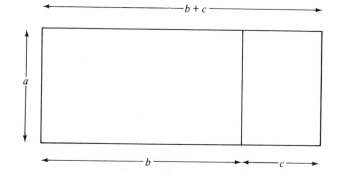

Use the figure to give a geometric interpretation of the equation

$$a(b + c) = ab + ac$$

13.

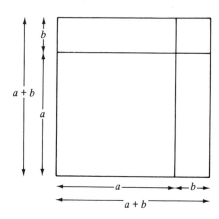

Use the figure to give a geometric interpretation of the equation

$$(a + b)^2 = a^2 + 2ab + b^2$$

14. Use the figure to give a geometric interpretation of the equation

$$(a + b)(a - b) = a^2 - b^2$$

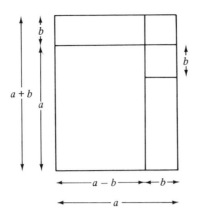

15. Find the area of a rectangle if $b = 2\frac{1}{2}$, $h = 1\frac{1}{4}$. Demonstrate by subdividing the rectangle into squares as in the text.

16. The perimeter of a rectangle is 20 ft. Find its length and its width if its area is to be made as large as possible. Use trial and error, computing the area for various choices of the length and the width.

3. Areas of Other Polygons

Using the formula for the area of a rectangle, we can now proceed to derive formulas for the areas of other polygons. Since a diagonal of a rectangle divides it into two congruent triangles (Chapter 6, Section 4) and since congruent triangles have the same area (Property 2), we conclude that each of the triangles must have area $\frac{1}{2}bh$. Thus for *right* triangles we have the formula, Area $= \frac{1}{2}bh$, where b and h are the lengths of the two perpendicular sides. (Figure 8-7)

Let us extend this formula to triangles in general. First we will define a *base of a triangle* to be any one of its sides. The *altitude corresponding to the base* is the segment which is perpendicular to the base and passes through the opposite vertex. Again "base" and "altitude" may be used to denote the lengths of these line segments. The length of an altitude of a triangle may also be called its *height*. Let b stand for the length of the base and h the length of the altitude corresponding to that base. There are three cases to consider. (Figure 8-8)

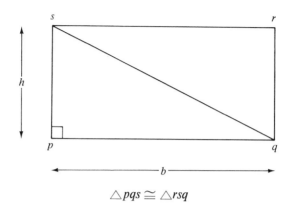

$$\triangle pqs \cong \triangle rsq$$

Figure 8-7

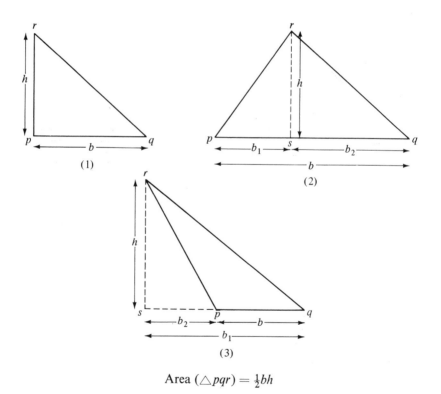

Area $(\triangle pqr) = \frac{1}{2}bh$

Figure 8-8

Case (1)

The altitude is one of the sides of the triangle. Then $\triangle pqr$ is a right triangle and we have already shown that Area $(\triangle pqr) = \frac{1}{2}bh$.

Case (2)

The altitude lies in the interior of the triangle, meeting the base in the point s. By the additive property

$$\text{Area } (\triangle pqr) = \text{Area } (\triangle psr) + \text{Area } (\triangle sqr)$$

Since $\triangle psr$ and $\triangle sqr$ are right triangles, their areas are $\frac{1}{2}b_1h$ and $\frac{1}{2}b_2h$ respectively, thus

$$\begin{aligned}\text{Area } (\triangle pqr) &= \tfrac{1}{2}b_1h + \tfrac{1}{2}b_2h \\ &= \tfrac{1}{2}(b_1 + b_2)h \\ &= \tfrac{1}{2}bh\end{aligned}$$

Case (3)

The altitude lies outside the triangle meeting the line containing the base in an exterior point s.

Again by the additive property

$$\text{Area } (\triangle sqr) = \text{Area } (\triangle spr) + \text{Area } (\triangle pqr)$$

Since $\triangle sqr$ and $\triangle spr$ are right triangles, their areas are $\frac{1}{2}b_1h$ and $\frac{1}{2}b_2h$, respectively. We then have

$$\tfrac{1}{2}b_1h = \tfrac{1}{2}b_2h + \text{Area } (\triangle pqr)$$

or

$$\begin{aligned}\text{Area } (\triangle pqr) &= \tfrac{1}{2}b_1h - \tfrac{1}{2}b_2h \\ &= \tfrac{1}{2}(b_1 - b_2)h\end{aligned}$$

Now $b_2 + b = b_1$, hence $b = b_1 - b_2$, and

$$\text{Area } (\triangle pqr) = \tfrac{1}{2}bh$$

If we can find the area of any triangle, then we can find the area of any polygon, because any polygonal region can be "triangulated," that is, broken up into a number of non-overlapping triangular regions whose union is the polygonal region. Thus in Figure 8-9 the area of the polygon is equal to the sum of the areas of triangles $T_1, T_2, T_3, T_4, T_5, T_6, T_7,$ and T_8.

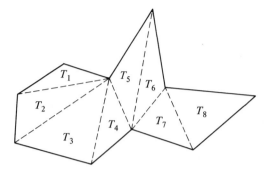

Triangulation

Figure 8-9

Now let us find a formula for the area of a parallelogram. Any side of a parallelogram is called a *base*. A line segment perpendicular to a base, one of whose end-points lies on the base and the other on the opposite side, is called an *altitude* to that base. There are infinitely many altitudes to a base, but since parallel lines are everywhere equidistant they all have the same length. In Figure 8-10 the segment $\overline{st}$ is an altitude to the base

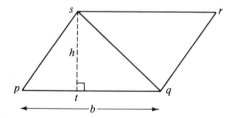

Area of a parallelogram = $b \cdot h$

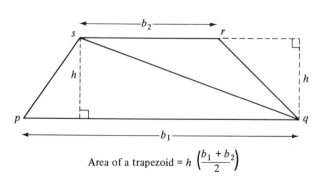

Area of a trapezoid = $h \left(\dfrac{b_1 + b_2}{2}\right)$

Figure 8-10

$\overline{pq}$. Let b stand for the length of the base and h the length of the corresponding altitude.

Since a diagonal divides the parallelogram into two congruent triangles,

$$\text{Area } (pqrs) = 2 \text{ Area } (\triangle pqs)$$
$$= 2(\tfrac{1}{2}bh)$$
$$= bh$$

The area of a parallelogram then is the product of the length of a base and the length of a corresponding altitude.

The formula for the area of a trapezoid is also found by triangulating the figure. Let $pqrs$ be a trapezoid with its two parallel sides (called its bases) having lengths b_1 and b_2.

$$\text{Area } (pqrs) = \text{Area } (\triangle pqs) + \text{Area } (\triangle qrs)$$

An altitude of a trapezoid is defined in the same way as the altitude of a parallelogram—a line segment perpendicular to the two bases whose end-points lie in these bases. The altitudes of both triangles are altitudes of the trapezoid, and they have the same length h, the distance between the parallel bases. Thus

$$\text{Area } (\triangle pqs) = \tfrac{1}{2}h \cdot b_1$$
$$\text{Area } (\triangle qrs) = \tfrac{1}{2}h \cdot b_2$$

and
$$\text{Area } (pqrs) = \tfrac{1}{2}h \cdot b_1 + \tfrac{1}{2}h \cdot b_2$$
$$= \tfrac{1}{2}h(b_1 + b_2)$$

We could rewrite this as $h\,(b_1 + b_2)/2$ and think of the area of a trapezoid as the product of the length of its altitude and the *average* of the lengths of its two bases. If the trapezoid is a parallelogram, then $b_1 = b_2$ and the formula becomes Area $= h \cdot b$.

Although the formula for the area of a triangle is very simple and easy to remember, it is not always practical to use. Suppose you want to find the area of a triangular plot of ground. Finding the length of a base is easy—you just use a tape measure—but if the triangle is not a right triangle, finding the altitude is not so simple. The formula below is known as *Heron's formula*. Heron, who lived in the first century B.C., was an inventor as well as a mathematician. He invented what must surely have been the first vending machine—a machine which sprinkled holy water when a coin was inserted! His formula gives the area of the triangle in terms of the lengths of its three sides, a, b, and c. Let $s = \tfrac{1}{2}(a + b + c)$. (This is half the perimeter.) Then the area of the triangle is given by the formula:

$$\text{Area} = \sqrt{s(s - a)(s - b)(s - c)}$$

Using Heron's formula, let us find the area of a triangle whose sides measure 5, 6, and 7 inches ($s = \frac{1}{2}(5 + 6 + 7) = 9$).

$$\text{Area} = \sqrt{9(9 - 5)(9 - 6)(9 - 7)}$$
$$= \sqrt{9 \cdot 4 \cdot 3 \cdot 2}$$
$$= 6\sqrt{6}$$

Now let us consider the problem of finding the area of a regular polygon. There are many ways to triangulate a regular polygon, but if we divide it into congruent triangles our work will be much easier.

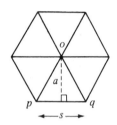

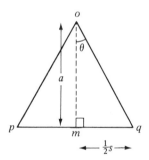

Finding the area of a regular hexagon

Figure 8-11

Suppose that we have a regular hexagon with side of length s. We can break it into six congruent triangles. Since these triangles are all congruent the area of the hexagon will be six times the area of one of them. Area $(\triangle pqo) = \frac{1}{2}a \cdot s$, where a is the length of the altitude to the side s. This length is called the *apothem* of the polygon. The area of the hexagon is therefore $6(\frac{1}{2}a \cdot s) = \frac{1}{2}a(6s)$ or $\frac{1}{2}a \cdot P$, where $P = 6s$ is the perimeter of the hexagon.

In general the area of *any* regular polygon will be given by the formula Area $= \frac{1}{2}aP$, where a is the apothem and P the perimeter of the polygon.

Since a regular hexagon is completely determined by the length of a side, if we know s we can find a by using trigonometry.

In Figure 8-11, $\triangle pqo$ is isosceles; therefore altitude $\overline{om}$ bisects side $\overline{pq}$ and $mq = \frac{1}{2}s$.

Since mqo is a right triangle

$$\tan \theta = \frac{\text{length of the opposite side}}{\text{length of the adjacent side}} = \frac{\frac{1}{2}s}{a}$$

(Chapter 7, Section 4)

Now θ is easy to find. There are six isosceles triangles, hence 12 right triangles in the hexagon, and they are all congruent. The angle has the same measure for all of them, and this measure must be $360/12 = 30$.

$$\tan 30 = \frac{1}{\sqrt{3}}$$

therefore

$$\frac{1}{\sqrt{3}} = \frac{\frac{1}{2}s}{a}$$

or

$$\frac{a}{\sqrt{3}} = \frac{1}{2}s$$

$$a = \frac{\sqrt{3}}{2}s$$

The area of the regular hexagon whose side is s is then given by the formula

$$\text{Area} = \frac{1}{2}\frac{\sqrt{3}}{2}s \cdot (6s)$$

$$= \frac{3\sqrt{3}}{2}s^2$$

Exercise 8.3

1. For a triangle, what would be the effect on

    (a) the area if its base were doubled and its altitude left unchanged?

    (b) the area if its base were doubled and its altitude cut in half?

    (c) the area if both its base and altitude were doubled?

    (d) the altitude if the area were doubled and the base were left unchanged?

    (e) the altitude if the area were doubled and the base cut in half?

    (f) the altitude if both the area and the base were doubled?

2. The ancient Egyptians used the incorrect formula Area =

$((a + c)/2)((b + d)/2)$ for finding the area of an arbitrary quadrilateral with successive sides having lengths a, b, c and d. Show that, in general, this formula does not hold. For what special cases of the quadrilateral will this formula hold?

3.   Sketch a number of parallelograms each having adjacent sides measuring 2 and 3 units. Explain why the parallelogram having maximum area is the one whose angles are right angles.

4.   Sketch a trapezoid with two non-parallel sides. Explain why this trapezoid will have maximum area when one of its angles is a right angle.

5.   Sketch a number of triangles each having two of its sides measuring 2 and 3 units. Explain why the triangle in which the two sides form a right angle has the largest area.

6.   An isosceles right triangle has area 32 sq. inches. What are the lengths of its sides?

7.   Find the area of the rhombus:

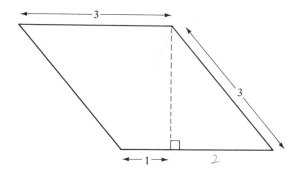

8.   In the triangle p, q and r are the midpoints of the sides. Find Area ($\triangle abc$); Area ($\triangle rqc$); Area ($\triangle apr$); Area ($\triangle pbq$); Area ($\triangle pqr$). (*Hint:* See Exercise 7.2, problem 15.)

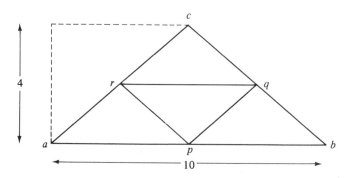

9. Find the areas of $\triangle abc_1$, $\triangle abc_2$, $\triangle abc_3$.

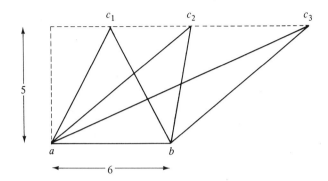

10. Find the area of the polygon *abcdef*.

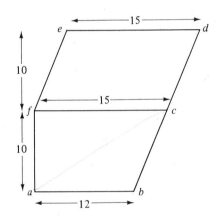

11. Find the area of polygon *abcde*.

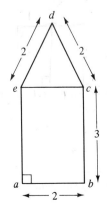

12. Show that the area of a right triangle is $\frac{1}{2}b \cdot h$ by showing that $\triangle pqu$ is congruent to $\triangle stu$ and hence

$$\text{Area } (\triangle prs) = \text{Area } ① + \text{Area } ②$$
$$= \text{Area } ③ + \text{Area } ②$$
$$= \tfrac{1}{2}bh$$

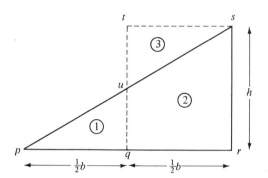

13. Prove that a median of a triangle divides it into two triangles of equal area. (A *median* of a triangle is a line segment joining a vertex to the midpoint of the opposite side.)

14. A triangle and a parallelogram have equal areas and equal bases. What can you say about their respective altitudes?

15. Suppose a quadrilateral has perpendicular diagonals. Prove that its area is equal to one-half the product of the lengths of the diagonals.

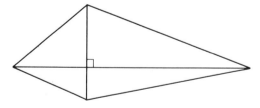

16. Use Heron's formula to show that the area of an equilateral triangle of side s is given by the formula $(\sqrt{3}/4)s^2$. Show that its altitudes must have measure $(\sqrt{3}/2)s$.

17. Find the area of an equilateral triangle with sides of length 4. What is the length of an altitude?

18. Suppose that we measures area in triangular units instead of square units. Let us define a *triangular unit* to be the area of an equilateral

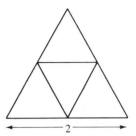

triangle each of whose sides measures one unit in length. Thus, for example, an equilateral triangle measuring two units on a side will have an area of 4 triangular units. What is the area in triangular units of an equilateral triangle whose side measures 3 units? 4 units? n units? Sketch. What is the area in *square* units of an equilateral triangle whose side measures 1 unit?

1 triangular unit = _____.

What is the area in triangular units of a square whose side measures 1 unit?

1 square unit = _____.

19. Prove or disprove the following statements:

   (a) If two polygons have the same perimeter then they have the same area.

   (b) If two polygons have the same area, then they have the same perimeter.

20. Derive the formula for the area of the trapezoid *pqrs* by computing the area of the rectangle *pqtu* and subtracting the areas of the triangles, $\triangle qtr$ and $\triangle psu$.

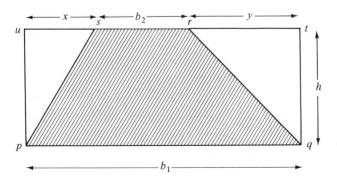

21.

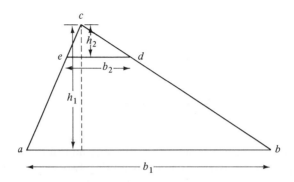

(a) Find the area of the trapezoid *abde*.
(b) Find Area ($\triangle abc$) — Area ($\triangle edc$).
(c) Use the fact that these two areas are the same to show that

$$\frac{b_1}{b_2} = \frac{h_1}{h_2}$$

22. Consider a regular hexagon whose side is one unit in length. (See Figure 8-11.)

(a) Show that each of the six isosceles triangles is equilateral. (*Hint:* Find the measure of the angles of the triangle.)

(b) Use Heron's formula to find the area of each equilateral triangle.

(c) Compute the area of the hexagon using (b) and compare the answer you get to that from using the formula given in this section.

23. Find the area of a regular polygon with five sides whose side measures *s* units, using the same method as in the text. You can find a value for tan θ from the table on p. 162.

24. Find the area of a regular polygon with 9 sides whose side measures *s* units.

25. Find the area of a regular polygon with ten sides whose side measures *s* units.

26. If the length of a side of a regular hexagon is doubled, what effect does this have on its area? What effect does this have on the apothem?

27. Any two regular hexagons are similar. Suppose the ratio of corresponding sides is 2. What is the ratio of their perimeters? What is the ratio of their areas?

4. The Pythagorean Theorem Revisited

There are hundreds of different proofs of the Pythagorean Theorem. We have already seen one which uses the properties of similar triangles in Chapter 7. One of the simplest proofs uses the properties of area. For simplicity we will change our notation and let A, B, and C stand for the vertices of the triangle and a, b, and c the lengths of the sides opposite these vertices.

Consider the right triangle $\triangle ABC$. We want to show that

$$a^2 + b^2 = c^2$$

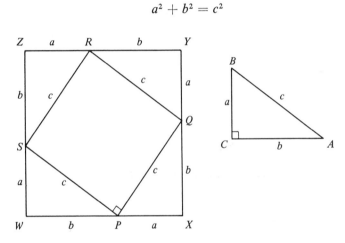

Proving the Pythagorean Theorem by using areas

Figure 8-12

First we construct a square WXYZ with sides of length $a + b$, and in each corner of the square construct a triangle congruent to $\triangle ABC$. (The triangles are congruent by the SAS Axiom.) Then each of the four triangles has the same area, $\frac{1}{2}ab$.

Since the two acute angles of a right triangle are complementary, $\angle SPQ$ must be a right angle, and so are the other three angles at Q, R, and S. $PQRS$ then must be a square, and its area is c^2.

By the additive property,

$$\text{Area } (WXYZ) = \text{Area } (PQRS) + 4 \text{ Area } (\triangle PXQ)$$

or
$$(a + b)^2 = c^2 + 4(\tfrac{1}{2}ab)$$

then $a^2 + 2ab + b^2 = c^2 + 2ab$

and $a^2 + b^2 = c^2$

Euclid's proof of this theorem also used areas, but was considerably more complicated than this one.

The Pythagorean Theorem can be stated in a purely geometric way. Given a right triangle ABC with sides of length a, b, and c, construct a square on each of the sides. The Pythagorean Theorem says that the area of the square on the hypotenuse is equal to the sum of the areas of the other two squares.

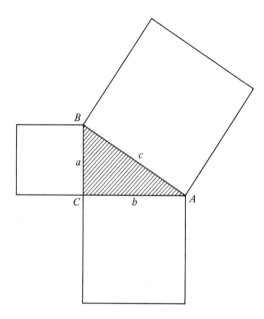

A geometric interpretation of the Pythagorean Theorem

Figure 8-13

This can be demonstrated by cutting apart the smaller squares and reassembling them to form the largest one.

Locate o, the intersection of the diagonals of the larger of the two squares on the legs of the triangle. (If they are the same size, then either one will do.) Through this center o, construct a line parallel to the hypotenuse of the triangle. Construct another line through o perpendicular to the first. If you cut out these four parts and the square on the other leg, they can be reassembled as indicated in Figure 8-14 to form the square on the hypotenuse. This, of course, does not constitute a proof. (See Chapter 6, Section 3)

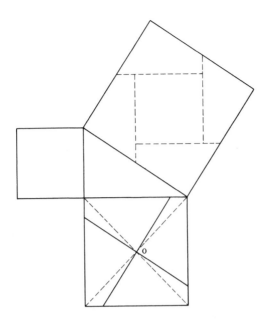

Demonstrating the Pythagorean Theorem by paper cutting

Figure 8-14

Exercise 8.4

1. The following proof of the Pythagorean Theorem is credited to General James A. Garfield, who later became President of the United States. Give reasons for each step in this proof.

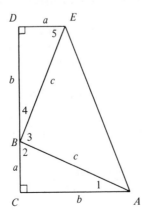

Given: $\triangle ABC$, $\angle C$ a right angle. Construct $\overline{BD}$ so that $BD = b$, and $\overline{DE} \perp \overline{BD}$ so that $DE = a$.

(a) $\overline{DE} \cong \overline{CB}$
(b) $\overline{DB} \cong \overline{CA}$
(c) $\angle C \cong \angle D$
(d) $\triangle ABC \cong \triangle BED$
(e) $\angle 1 \cong \angle 4$
(f) $\angle 1$ and $\angle 2$ are complementary
(g) $\angle 2$ and $\angle 4$ are complementary
(h) $\angle 3$ is a right angle
(i) $\triangle AEB$ is a right triangle
(j) $AEDC$ is a trapezoid
(k)   Area $(AEDC) = \frac{1}{2}(a + b)(a + b)$
(l)   Area $(\triangle ABC) = \frac{1}{2}ab$
(m)   Area $(\triangle BDE) = \frac{1}{2}ab$
(n)   Area $(\triangle AEB) = \frac{1}{2}c^2$
(o)   Area $(AEDC) =$ Area $(\triangle ABC) +$ Area $(\triangle BED) +$ Area $(\triangle AEB)$, i.e., $\frac{1}{2}(a + b)(a + b) = \frac{1}{2}ab + \frac{1}{2}ab + \frac{1}{2}c^2$
(p) $a^2 + b^2 = c^2$

2.   Use the Pythagorean Theorem to find the altitude of an isosceles triangle if its base has measure 12 and the congruent sides each have measure 10. Use this to compute its area. Check your answer by using Heron's formula.

3.   Show that if the congruent sides of an isosceles right triangle have measure a, then the hypotenuse has measure $\sqrt{2}\,a$. Find the length of the diagonal of a square with sides of length 5.

5. Areas of Similar Polygons

Suppose that you have two similar rectangles $pqrs$ and $p'q'r's'$, and each side of $p'q'r's'$ is twice as long as the corresponding side of $pqrs$. (Figure 8-15) The ratios of the two areas will be

$$\frac{\text{Area } (p'q'r's')}{\text{Area } (pqrs)} = \frac{(2b)(2h)}{bh} = 2 \cdot 2 = 4$$

To be more general, if each side of $p'q'r's'$ is k times as long as the corresponding side of $pqrs$, then the ratio of the two areas will be

$$\frac{\text{Area } (p'q'r's')}{\text{Area } (pqrs)} = \frac{(kb)(kh)}{bh} = k \cdot k = k^2$$

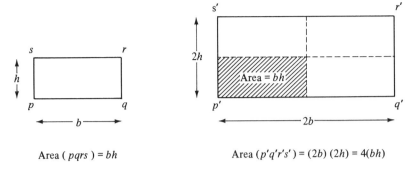

Area (*pqrs*) = *bh*   Area (*p′q′r′s′*) = (2*b*) (2*h*) = 4(*bh*)

Figure 8-15

Thus if the ratio of the corresponding sides is 3, the area of the second will be 9 times the area of the first; if the ratio of corresponding sides is $\frac{1}{2}$, the ratio of the areas will be $\frac{1}{4}$.

The same statement can be made for the ratio of the areas of two similar triangles. If we have two similar triangles, $\triangle pqr$ and $\triangle p'q'r'$, and each side

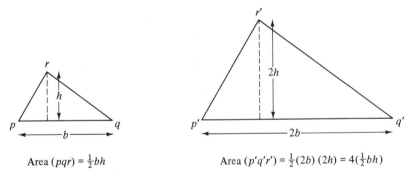

Area (*pqr*) = $\frac{1}{2}bh$ Area (*p′q′r′*) = $\frac{1}{2}$(2*b*) (2*h*) = 4($\frac{1}{2}bh$)

Figure 8-16

of $\triangle p'q'r'$ is twice as long as the corresponding side of $\triangle pqr$, then each altitude of $\triangle p'q'r'$ will be twice as long as the corresponding altitude of $\triangle pqr$. The ratio of the two areas then will be

$$\frac{\text{Area } (\triangle p'q'r')}{\text{Area } (\triangle pqr)} = \frac{\frac{1}{2}(2b)(2h)}{\frac{1}{2}bh} = 2 \cdot 2 = 4$$

or, in general, if the ratio of corresponding sides is *k*, the ratio of the areas is

$$\frac{\text{Area } (\triangle p'q'r')}{\text{Area } (\triangle pqr)} = \frac{\frac{1}{2}(kb)(kh)}{\frac{1}{2}bh} = k^2$$

This statement will be true for any pair of similar polygons. Each of the formulas we have developed for the areas of polygons involves the product of two lengths.

$$\text{Area of a Parallelogram} = bh$$
$$\text{Area of a Square} \quad = s^2 \text{ or } s \cdot s$$
$$\text{Area of a Trapezoid} \quad = h \left(\frac{b_1 + b_2}{2} \right)$$

(Here $(b_1 + b_2)/2$ is the *average* length of the two bases.)

$$\text{Area of a Regular Hexagon} = \frac{3\sqrt{3}}{2} s^2$$

If the length of a line segment in one polygon is k times the length of the corresponding segment in a similar polygon, then the ratio of the two areas will be k^2.

6. Areas of Irregular Regions

All of the formulas for area that we have developed have been for polygons. What about regions with curved boundaries? They have areas, but how can we find them? To find such areas exactly is a difficult problem, solved (theoretically, at least) in calculus. We can, however, find such areas approximately, using the theory of area developed in this chapter, and for practical purposes this may be all that is desired.

To approximate the area of the region R, we will use a *grid* or *mesh*. A grid is made up of a number of vertical and horizontal lines, at regularly-spaced intervals. In a one-inch grid, for example, the lines are one inch apart. Each square in the grid has area 1 square inch.

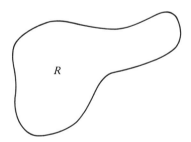

The area of the region R can be found approximately by using a grid.

Figure 8-17

If we superimpose a one-inch grid over the figure, then some of the squares will lie completely in the interior of the region *R*. We have shaded these in Figure 8-18—there are 4 of them. We can say then that the area of R is *greater than* 4 square inches.

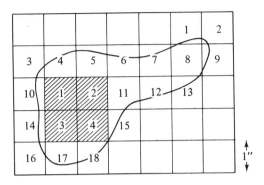

4 sq. in. < Area *R* < 22 sq. in.

Figure 8-18

There are other squares which are partly in the interior of the figure and partly in its exterior. The boundary cuts through these squares—there are 18 of them. The entire region *R* lies in the polygonal region formed by taking the union of the interior squares and the boundary squares. This exterior polygon has area $18 + 4 = 22$ square inches. Clearly, the area we seek is *less than* 22 square inches

We conclude that

4 square inches < Area of R < 22 square inches

Now you may object that this is not a very close approximation. To get a better approximation, we can use a finer grid—say a $\frac{1}{2}''$ grid. In a $\frac{1}{2}''$ grid, the lines are $\frac{1}{2}''$ apart, and each square has area $\frac{1}{4}$ square inch.

Using this finer grid, we count 32 squares in the interior polygon or $\frac{32}{4} = 8$ square inches, and $35 + 32 = 67$ in the exterior polygon, or $\frac{67}{4} = 16\frac{3}{4}$ square inches. (Figure 8-19) We have then

8 square inches < Area of *R* < $16\frac{3}{4}$ square inches

A fairly good approximation to the area would be the average of the two, or 12.4 square inches. If we wanted a still more accurate figure, we could use an even finer grid. You can see that we could, by using finer and finer grids, find the area to any desired degree of accuracy.

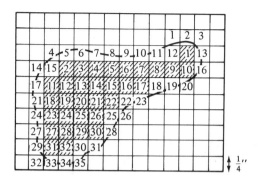

8 sq. in. $<$ Area $R <$ 16 $\frac{3}{4}$ sq. in.

Figure 8-19

Exercise 8.6

1. The ratio of corresponding sides of two similar polygons is $\frac{5}{3}$. What is the ratio of their areas? The ratio of the areas of two similar polygons is 25. What is the ratio of corresponding sides?

2. Let *abcd* be a trapezoid with $b_1 = 5$, $b_2 = 7$, $h = 6$, and let *a'b'c'd'* be a similar trapezoid whose sides are $\frac{1}{2}$ as long. Verify directly by computing the areas that the ratio of the areas is $(\frac{1}{2})^2 = \frac{1}{4}$.

3. A rectangle has sides of length 2 and 3 units. A similar rectangle has one side of length 12 units. What is the area of the second rectangle? (There are two answers to this problem.)

4. Two similar triangles have area 16 and 25 square units, respectively. If one side of the smaller triangle has length 8 units, how long is the corresponding side of the larger triangle?

5. Draw a circle with radius 3 inches. Approximate its area by using a one-inch grid; by using a $\frac{1}{2}$-inch grid. The exact area of this circle is 9π. What approximate value do you get for π?

6. Draw an isosceles right triangle with legs 3 inches. Approximate its area by using a one-inch grid; by using a $\frac{1}{2}$-inch grid. Compare the approximation you get with the exact area obtained from the formula.

New Terms Found in This Chapter

Term	Section	Term	Section
additive property of area	2	height of a triangle	3
square unit	2	base of a parallelogram	3
base of a rectangle	2	altitude of a parallelogram	3
altitude of a rectangle	2	base of a trapezoid	3
base of a triangle	3	altitude of a trapezoid	3
altitude of a triangle	3	triangular unit	Exercise 8.3
triangulation	3	grid or mesh	6

9: Space Figures and Volumes

1. Regular Polyhedra

In an earlier chapter we constructed a plane figure, the polygon, from line segments. A polygonal region was defined to be the union of a polygon and its interior. We now move into solid geometry by constructing a space figure, called a *polyhedron* (plural, *polyhedra*) from polygonal regions.

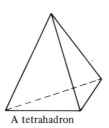
A tetrahadron

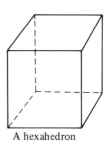
A hexahedron

Figure 9-1

The polygonal regions are called the *faces* of the polyhedron. A polyhedron with four faces is called a tetrahedron, one with six faces a hexahedron, with eight faces an octahedron, and so on. The intersection of two faces of a polyhedron is either a line segment, a single point or the empty set. The line segments are called the *edges* of the polyhedron and the end points of the edges are called its *vertices*. Consider, for example, the tetrahedron in Figure 9-1. It has four faces, three lateral sides and a base. These meet in 6 edges, which in turn intersect in 4 vertices. The hexahedron in the figure has 6 faces, 12 edges, and 8 vertices. The faces of a polyhedron may be triangles, quadrilaterals, or any other polygons, or any combination of these.

The polyhedra we have described are hollow. The union of the polyhedron and its interior will be called a *solid polyhedron*. The reader should be cautioned that there is much disagreement among geometers on these terms. He may find the word "polyhedron" used as we have used it to mean the surface only, or he may find it used to mean the solid.

In this section all of the solid polyhedra we study will be *convex*. This means that given any two points in the solid, the line segment joining them will lie completely within the solid. This will rule out consideration of such a solid polyhedron as a box with a dent in it.

A polyhedron is *regular* if all of its faces are congruent regular polygons. Recall that a polygon is regular if all of its sides are congruent and all of its angles are congruent as well. The regular triangle is an equilateral triangle, and the regular quadrilateral is a square. We can have regular polygons with any number, n, of sides ($n \geq 3$); however, regular polyhedra are not so common. There are only five regular polyhedra, those having four, six, eight, twelve, and twenty faces.

A regular tetrahedron has four faces, each of which is an equilateral triangle.

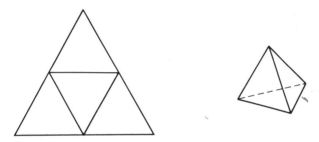

A pattern for a regular tetrahedron

Figure 9-2

Each face of a regular hexahedron is a square. This is commonly called a *cube*.

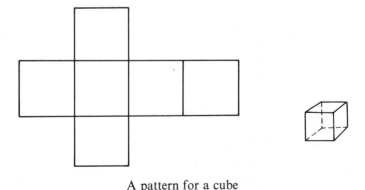

A pattern for a cube

Figure 9-3

A regular octahedron has eight faces and each face is an equilateral triangle.

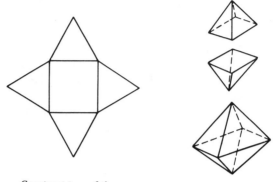

Construct two of these.

A pattern for a regular octahedron

Figure 9-4

A regular dodecahedron has twelve faces, each of which is a regular pentagon.

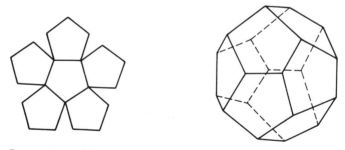

Construct two of these.

A pattern for a regular dodecahedron.

Figure 9-5

The Greeks admired the regular polyhedra so much that they believed that all objects were made up of fundamental particles having these shapes. They believed that there were four elements, earth, air, fire, and water. According to Plato, the fundamental particles of fire were tetra-

A regular icosahedron has twenty faces, and each face is an equilateral triangle.

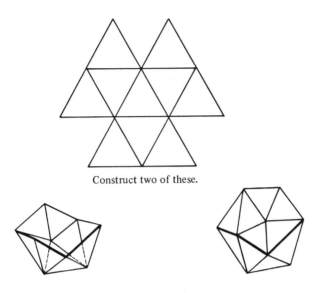

Construct two of these.

A pattern for a regular icosahedron

Figure 9-6

hedra, those of air, octahedra, those of water, icosahedra, and of earth, cubes. The fifth shape, the dodecahedron, was the shape of the universe itself.

Actually the five regular polyhedra *do* occur in nature, though not as particles of fire, water or earth. The tetrahedron, the cube, and the octahedron occur as crystals. Common salt, for example, occurs in crystals which are cubes. The crystals of sodium bromate are tetrahedra, and those of chrome alum are octahedra. The dodecahedron and icosahedron do not occur as crystals, but skeletons of microscopic sea animals called *Radiolaria* have been found to have these forms.

Why are there only five regular polyhedra? How do we know that there are not more we simply haven't thought of yet? To begin with, let us note two facts. First, in any polyhedra at *least* three faces meet at each vertex. There may be more than three, but there cannot be fewer. Second, if we take the sum of the measures of all the angles having a common vertex, this sum must be *less than* 360. If the sum were equal to 360, the surface would be flat at this point and we would not call the point a vertex. If the sum were greater than 360, then the solid would not be convex.

Now let us consider all possible regular polyhedra whose faces are equilateral triangles. Each angle of an equilateral triangle has measure 60. Since the angle sum at each vertex must be less than 360, there could be 3, 4, or 5 faces meeting at each vertex ($3 \cdot 60 = 180$; $4 \cdot 60 = 240$; $5 \cdot 60 = 300$). Six faces (or more) could not meet at one vertex, however, since $6 \cdot 60 = 360$. These three cases correspond to the regular tetrahedron, the regular octahedron and the regular icosahedron. These are the only three regular polyhedra having triangular faces.

Next we look at the regular polyhedra whose faces are squares. Since each angle of a square measures 90, three faces could meet at each vertex ($3 \cdot 90 = 270$), but not four or more ($4 \cdot 90 = 360$). This, of course, describes the cube.

Each angle of a regular pentagon measures $3 \cdot 180/5 = 108$. (Exercise 6.3, problem 16) A regular polyhedron whose faces are pentagons can therefore have at most three faces meet at each vertex ($3 \cdot 108 = 324$). This case corresponds to the regular dodecahedron.

What about a regular polyhedron whose faces are hexagons, or heptagons? They cannot exist. Each angle of regular hexagon measures $4 \cdot 180/6 = 120$, and $3 \cdot 120 = 360$, hence three such faces could not meet at one vertex. Since the angle measure of regular polygons gets larger as the number of sides increases, it is not hard to see that regular polyhedra whose faces are polygons with more than five sides cannot exist.

2. Pyramids and Prisms

Two special types of solid polyhedra are the pyramid and the prism. Before we describe these space figures, let us briefly review some facts about planes and lines in space. Two planes are parallel if their intersection is empty. If a line intersects a plane, it either lies in the plane, or it meets the plane in exactly one point. If a line intersects a plane in exactly one point, it will intersect any parallel plane in exactly one point also.

We say that a line L is perpendicular to a plane Π if it intersects the plane in a point q, and is perpendicular to every line in the plane which passes through q. The distance from a point p to a plane Π is the perpendicular distance, (the length of a line segment $\overline{pq}$, where $q \in \Pi$ and $\overline{pq}$ is perpendicular to Π). The distance between two parallel planes Π_1 and Π_2 is the distance from any point p on Π_1 to Π_2. (Figure 9-8)

Now let Π be a plane and p any point not in Π. Let A be a polygonal region in Π. Then the union of all the line segments $\overline{pa}$, where $a \in A$ is called a *solid pyramid*, or simply a *pyramid*, with base A and vertex p. (Figure 9-9) If q is in Π and $\overline{pq}$ is perpendicular to Π, we will call the line segment $\overline{pq}$ the *altitude* of the pyramid. The *length* of $\overline{pq}$ is also called the altitude or the *height* of the pyramid. The faces of the pyramid which are

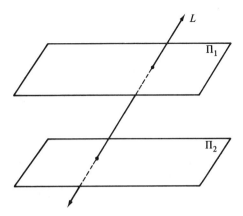

Planes Π_1 and Π_2 are parallel. Since L intersects Π_1 in exactly one point, it intersects Π_2 in exactly one point also.

Figure 9-7

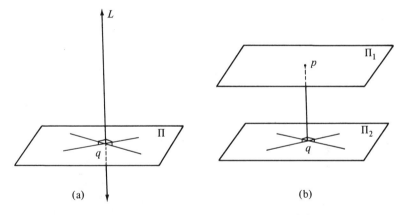

(a)   (b)

(a)   Line L is perpendicular to plane Π since it is perpendicular to every line in Π passing through q.

(b)   The distance between the parallel planes Π_1 and Π_2 is pq.

Figure 9-8

not in the plane Π are called its *lateral faces*. Every lateral face of a pyramid will be a triangular region. The base, of course, may be any polygonal region.

The intersection of a plane parallel to Π and the pyramid, if this intersection is not empty, is called a *horizontal cross section* of the pyramid. This cross section will be either a single point or it will be a polygonal

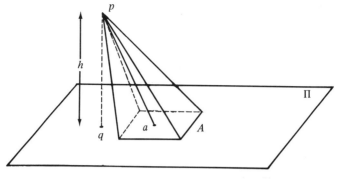

A pyramid with base A, vertex p, and height pq

Figure 9-9

region similar to the base. By "similar to" we mean mathematical similarity, as defined in Chapter 7.

If instead of a polygonal region in the plane we have some region bounded by a simple closed curve C, the solid formed is called a *solid cone*. If C is a circle, then it is called a (solid) *circular cone*.

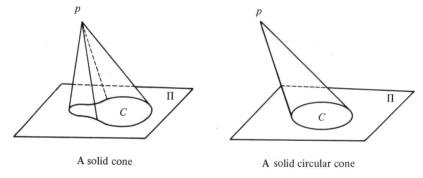

A solid cone   A solid circular cone

Figure 9-10

If the vertex p lies directly above the center of the circle, then we call this a (solid) *right circular cone*. In mathematics books when the reader finds the word "cone," usually it is a right circular cone that is meant.

Now consider two parallel planes, Π_1 and Π_2. Let L be a line that intersects both of them, and let A_1 be a polygonal region in Π_1.

For each point p_1 in A_1, consider the line segment $\overline{p_1 p_2}$ parallel to L, where $p_2 \in \Pi_2$. The union of all of these line segments is called a (solid) *prism*. A_1 is called the *lower base*, and A_2, the intersection of the prism with the plane Π_2, is called the *upper base*. The faces which are not in Π_1 or Π_2 are called the *lateral faces* of the prism. They are all parallelogram

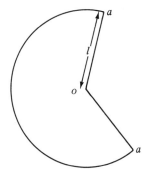 

A pattern for a right circular cone. The length l is called the *slant height*.

Figure 9-11

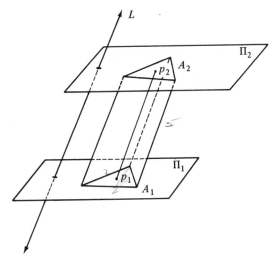

A prism with lower base A_1 and upper base A_2

Figure 9-12

regions. It is not hard to see why this is so. Two of the opposite sides lie in the parallel planes Π_1 and Π_2 and the other two sides are both parallel to L, hence are parallel to each other.

Since opposite sides of a parallelogram are congruent, corresponding sides of A_1 and A_2 are congruent, and if A_1 and A_2 are triangles, it follows that they are congruent by the SSS Axiom.

The two bases A_1 and A_2 will always be congruent, even if they are polygons other than the triangle. Suppose, for example, A_1 and A_2 are quadrilaterals (Figure 9-13). Since the corresponding diagonals of A_1 and A_2 are opposite sides of a parallelogram, they are congruent, and each of the two triangles in A_1 is congruent to the corresponding triangles in A_2, again by the SSS Axiom. Thus A_1 is congruent to A_2. Since any polygon can be broken up into triangles, this same reasoning leads us to the conclusion that the bases of a prism are always congruent. For the same reasons, any horizontal cross section of a prism will be congruent to the bases.

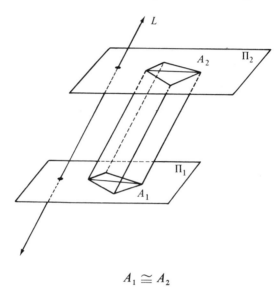

$$A_1 \cong A_2$$

Figure 9-13

If the base of the prism is a parallelogram region, then the prism is called a *parallelepiped*. All of the faces of a parallelpiped are parallelogram regions.

The distance between the two planes Π_1 and Π_2 is called the *altitude* or the *height* of the prism. If the line L is perpendicular to the planes, then the solid is called a *right prism*. A right prism whose base is a rectangle is a *rectangular parallelopiped*. Its surface, a hollow prism, is called a *box*.

A more general kind of solid which includes the prism as a special case is the *cylinder*. Again, let Π_1 and Π_2 be parallel planes, L a line cutting both of them. The base of the cylinder may be any region whose boundary is a simple closed curve. Since a polygon is a simple closed curve, a prism is a special kind of cylinder. (Figure 9-14)

If the base is a circular region, then the cylinder is called a *circular cylinder*. If L is perpendicular to the two planes, it is called a *right circular cylinder*. (Figure 9-15)

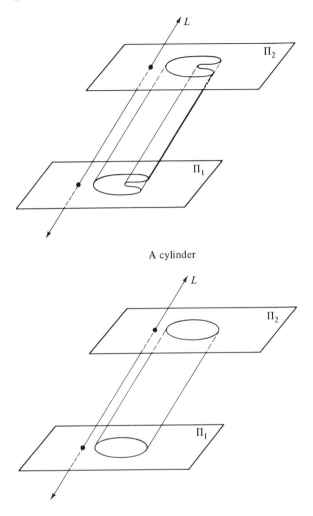

A cylinder

A circular cylinder

Figure 9-14

Note that cylinders, prisms, and pyramids are all solid figures the way we have defined them. If we wish to discuss the surface only, we will refer to the "surface of a prism" or a "hollow prism."

Another common space figure is the *sphere*. A *sphere* is the set of all points in space at a fixed distance r from some fixed point p. This definition differs from the definition of a circle by the substitution of the phrase "in space," for "in a plane." (Chapter 3, Section 3) The point p is called

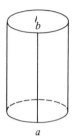

A pattern for a right circular cylinder

Figure 9-15

the center of the sphere and *r* is called its radius; *r* is the measure of any line segment $\overline{pq}$ where *q* is a point on the sphere. The segment $\overline{pq}$ is also called a radius of the sphere. A diameter of a sphere is any line segment $\overline{qs}$ which contains *p*, where *q* and *s* are points on the sphere. The length of a diameter is twice the length of a radius. We also refer to this number as the diameter of the sphere.

The sphere is hollow. If we wish to refer to the union of the sphere and its interior we call this a *solid sphere* or a *ball*.

Every cross section of a sphere is a circle (or a single point). If a plane passes through *p*, then the intersection of the plane and the sphere is called a *great circle*. Note that a great circle is a subset of the sphere, but the radius and diameter are not.

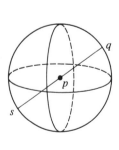

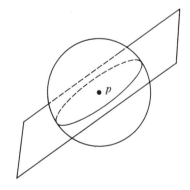

The sphere with center *p*,
radius *pq*, and diameter *sq*.

A great circle

Figure 9-16

Exercise 9.2

1. Name a common household object that seems to be a model of a pyramid; a circular cone; a circular cylinder; a prism; a sphere; a cube. We speak of an ice "cube" or a "cube" of butter. Are these cubes as we have defined the term?

2. What is the smallest number of faces that a polyhedron may have? What is the smallest number of edges? The smallest number of vertices?

3. Using heavy paper or cardboard construct a model of (a) a regular tetrahedron; (b) a cube; (c) a regular octahedron; (d) a regular dodecahedron; (e) a regular icosahedron.

4. Complete the following table:

	F No. of Faces	E No. of Edges	V No. of Vertices
regular tetrahedron			
cube			
regular octahedron			
regular dodecahedron			
regular icosahedron			

Compare $F + V$ with E. What do you conclude?

5. Complete the following table for a pyramid whose base is given in the first column:

Base	F No. of Faces	E No. of Edges	V No. of Vertices
triangle			
quadrilateral			
pentagon			
hexagon			
.			
.			
.			
n-gon			

In each case compare $F + V$ with E. What do you conclude?

6. Complete the following table for a prism whose base is given in the first column:

Base	F No. of Faces	E No. of Edges	V No. of Vertices
triangle			
quadrilateral			
pentagon			
hexagon			
.			
.			
.			
n-gon			

In each case compare $F + V$ with E. What do you conclude?

7. Prove that if there is a correspondence between two quadrilaterals such that four corresponding sides and a corresponding diagonal are congruent, then the two quadrilaterals are congruent. (Recall that two polygons are congruent if there is a one-to-one correspondence between their vertices such that corresponding sides and corresponding angles are congruent.)

8. Draw a Venn diagram showing the relationship between the set of solid polyhedra, the set of regular solid polyhedra, the set of prisms and the set of pyramids. Where does the solid cube fit into this picture? The solid tetrahedron?

9. Draw a Venn diagram showing the relationship between the set of (solid) cylinders; the set of solid polyhedra and the set of prisms. Where does the (solid) right circular cylinder fit into this picture?

10. The *surface area* of a polyhedron is the sum of the areas of its faces. Find the surface area of a cube each of whose edges has length 2.

11. Find the surface area of a regular tetrahedron, each of whose edges has length 1.

12. Find the surface area of a regular octahedron, each of whose edges has length 3.

13. Find the surface area of a regular icosahedron, each of whose edges has length 2.

14. Develop a formula for the surface area of a regular tetrahedron each of whose edges has length s. Also develop a formula for the surface area of a regular octahedron, and of a regular icosahedron.

15. Find the surface area of a right prism whose base is an equilateral triangle with sides of length 2 and whose height is 5.

16. Develop a formula for the surface area of a rectangular parallelepiped

letting l and w stand for the length and width of the base, and h stand for the height. What is the formula for the surface area of the cube?

17. The intersection of a plane and a sphere might be (a) a great circle; (b) some circle which is not a great circle; (c) a single point. Sketch each of these three cases.

3. Euler's Formula

A formula relating the number of faces, edges, and vertices of a simple polyhedron was discovered originally by Descartes, and rediscovered and used by Euler in 1752. (A polyhedron is "simple" if there are no "holes" in it. All of the polyhedra we have discussed in this chapter are simple.)

If V stands for the number of vertices, E the number of edges, and F the number of faces, then

$$V - E + F = 2$$

For a simple proof of Euler's formula, the reader is referred to Courant and Robbins, *What is Mathematics*; 4th ed. (New York: Oxford University Press, Inc., 1941), pp. 236–240.

Exercise 9.3

1.

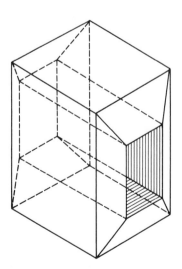

The polyhedron above is not simple. It is a box with a hole in it. Show that Euler's formula does not hold for this surface.

2. If three triangular faces meet at each vertex of a regular polyhedron, then there are $3V$ face angles; $3V/3 = V$ faces, since each face has three face angles, and $3V/2$ edges, since each face is bound by 3 edges and each edge is an edge of two faces. Complete the following table:

	No. of Face Angles (in terms of V)	No. of Faces (in terms of V)	No. of Edges (in terms of V)
(1) 3 triangular faces meet at each vertex	$3V$	V	$3V/2$
(2) 3 square faces meet at each vertex		$3V/4$	
(3) 4 triangular faces meet at each vertex	$4V$		
(4) 3 pentagonal faces meet at each vertex		$3V/5$	
(5) 5 triangular faces meet at each vertex			$5V/2$

3. According to the table above, a polyhedron whose faces are equilateral triangles, 3 of which meet at each vertex has V faces and $3V/2$ edges. Substituting these for F and E in Euler's formula, we get

$$V - \frac{3V}{2} + V = 2 \qquad \text{or } V = 4$$

But $V = F$, hence this polyhedron must be the regular tetrahedron. Do this for each of the other rows in the table, thus showing that for row (2) $F = 6$; for row (3) $F = 8$; for row (4) $F = 12$; and for row (5) $F = 20$.

4.   Some Basic Assumptions about Volume

Volume is a measure in space—in three dimensions, just as area is a measure in the plane (two dimensions), and length is a measure on the line (one dimension).

Volumes are measured in terms of *cubic units*. A cubic inch, for example, is defined to be the volume of a cube measuring one inch on each edge.

A cubic foot or a cubic yard can be defined in the same way. Since a cubic yard is the volume of a cube measuring 1 yard on each edge, and

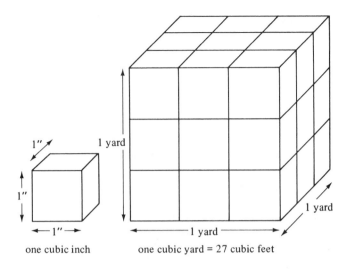

one cubic inch   one cubic yard = 27 cubic feet

Figure 9-17

there are three feet in a yard (linear measure), then twenty-seven cubic feet
can fit into a cubic yard.

Intuitively we can think of volume measurement as finding the number
of cubic units that will fit snugly into the solid being measured. If a rectangular parallelepiped (a solid box) has a base which is 3 units by 4 units
and height 2 units, then we can think of fitting in 24 cubic units—two
layers with $3 \cdot 4 = 12$ cubic units in each layer.

Experimentally, if we have a hollow space figure, then the volume
of the solid can be found by filling it with water or sand, then measuring
the volume of the water or sand.

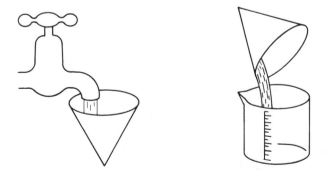

Measuring the volume of a cone experimentally

Figure 9-18

More formally we find the volume of certain solids by using formulas. In this chapter we will develop formulas for finding the volume of some common polyhedra.

We will assume that every solid polyhedron has associated with it a positive real number called its *volume*. If A is a solid, then the symbol Vol (A) will denote its volume. Some of the basic properties of volume measurement are:

(1) The volume of a solid polyhedron is a positive real number.
(2) If A and B are two solid polyhedra and $A \subset B$, then Vol $(A) \leq$ Vol (B).
(3) If A and B are two solid polyhedra whose intersection is (a) empty; (b) a single point; (c) a line segment; or (d) a region in the plane, then Vol $(A \cup B) =$ Vol $(A) +$ Vol (B).

This last property is called the additive property of volume.

We will start out by assuming that the volume of a rectangular parallel-epiped (a solid box) is the product of the area of its base and its height. Since the base of a box is a rectangle, its area is the product of the lengths of two adjacent sides. Thus we have our first formula for finding volumes. If A is a rectangular parallelepiped, then

$$\text{Vol } (A) = l \cdot w \cdot h$$

where l, w and h stand for its length, width, and height.

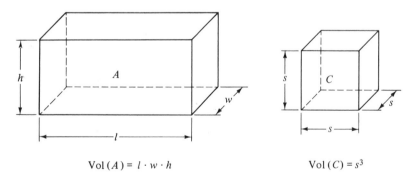

Vol $(A) = l \cdot w \cdot h$ Vol $(C) = s^3$

Figure 9-19

Using this formula we can proceed to derive formulas for volumes of some other polyhedra. A cube is a rectangular parallelepiped all of whose edges have the same measure, call it s. The volume of a cube C is then

$$\text{Vol } (C) = s \cdot s \cdot s = s^3$$

Read "s^3" as "s cubed." You can now see why this term is used for the product $s \cdot s \cdot s$.

5. Cavalieri's Principle *omit*

A theorem which is extremely useful in computing volumes is due to an Italian mathematician named Bonaventura Cavalieri (1598–1647), a pupil of Galileo. We shall not prove this theorem here but will state it as *Cavalieri's Principle*. This principle tells us, for example, that if two pyramids have congruent bases and have the same height, then they have the same volume even though their shape might be quite different (Figure

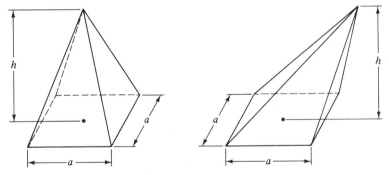

These two pyramids have the same volume.

Figure 9-20

9-20). This seems reasonable since we recall that a similar statement holds about the area of triangles. If two triangles have the same base and the same height, then they have the same area (Figure 9-21).

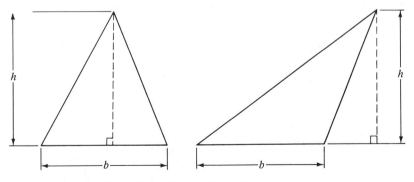

These two triangles have the same area.

Figure 9-21

Actually Cavalieri's Principle tells us much more than this. Roughly speaking, this principle states that if two solids have the same cross sectional area at all levels, then they have the same volume, even though their shapes are different. Let us state this principle more precisely.

Cavalieri's Principle

Let A_1 and A_2 be two solids which lie between two parallel planes Π_1 and Π_2. Let Σ be any plane parallel to Π_1 and Π_2 and let K_1 and K_2 denote the cross sections of A_1 and A_2 determined by this plane. If for *every* such plane Σ the cross sections K_1 and K_2 have the same area, then A_1 and A_2 have the same volume.

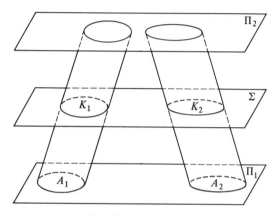

Cavalieri's Principle

Figure 9-22

We can illustrate this principle with two identical decks of playing cards. Stacked in the usual way, the first deck looks like a solid box, and its volume can be easily calculated. Now suppose we take the second deck and push the stack to one side so that it resembles Figure 9-23. The cross sections determined by any plane parallel to the base of these two solids will have the same area, since any two cards are congruent. Clearly, the two decks have the same volume, since they were identical.

The cross sections do not have to be congruent, they only need to have the same area. For example, consider the two prisms in Figure 9-24.

One has a rectangular base with sides measuring 3 and 2 units, and the other has a right triangle for its base with legs measuring 6 and 2 units. They have the same height—12 units. Since the two bases have the same area—6 square units, and every horizontal cross section of a prism is congruent to the base (Section 2), by Cavalieri's Principle, the two prisms have the same volume—72 cubic units.

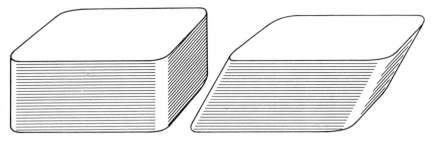

Illustrating Cavalieri's Principle with two decks of playing cards

Figure 9-23

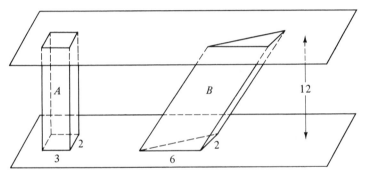

Vol (A) = Vol (B) = 72 cubic units.

Figure 9-24

6. The Volume of a Prism

We can now use the basic assumptions about volume and Cavalieri's Principle to derive a formula for the volume of a prism.

We have assumed in Section 4 that the volume of a rectangular parallelepiped (a solid box) is the product of the area of its base and its height. Now consider a right prism P_1 whose base is a triangle. (Figure 9-25)

The area of the base is $\frac{1}{2}a \cdot b$. We can construct a solid box whose base is a rectangle having the same area—say one whose length is b and whose width is $\frac{1}{2}a$. Assume that the box and the triangular prism have the same height h. Since every horizontal cross section of these solids is congruent to the base, any plane parallel to the base will determine cross sections of these two solids which have the same area, and the two solids will have the same volume. The volume of the solid box we can find by

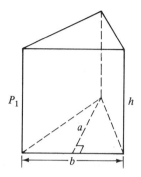 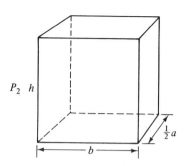

$$\text{Vol } (P_1) = \text{Vol } (P_2) = (\tfrac{1}{2}ab) \cdot h$$

Figure 9-25

our basic assumption about volume. It is equal to the area of its base multiplied by its height.

$$\text{Vol } (P_2) = (\tfrac{1}{2}a \cdot b) \cdot h$$

Since $\text{Vol } (P_2) = \text{Vol } (P_1)$ and the area of the base of P_2 is the same as the area of the base of P_1, we have

$$\text{Vol } (P_1) = \text{Area of the base} \cdot \text{height}$$

This argument can easily be extended to a right prism whose base is *any* polygonal region, *B*. We can construct a solid box having the same height *h*, and whose base is a rectangle whose area is the same as the area of *B*. A rectangle whose length is numerically equal to the area of *B* and whose width is one unit would do. By Cavalieri's Principle these two solids have the same volume—the area of the base times the height. We thus arrive at a general formula for the volume of any right prism P_r.

$$\text{Vol } (P_r) = \text{Area of the base} \cdot \text{height}$$

But why stop here? Cavalieri's Principle also applies to prisms which are not necessarily right prisms.

Let *P* be any prism whatever. Since any horizontal cross section of *P* is congruent to the base, by Cavalieri's Principle its volume is equal to the volume of a solid box having the same height as *P* and whose base has the same area as the base of *P*. Thus we have a formula for the volume of any prism *P*. (Figure 9-26)

$$\text{Vol } (P) = \text{Area of the base} \cdot \text{height}$$

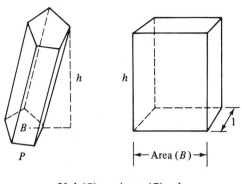

$$\text{Vol } (P) = \text{Area } (B) \cdot h$$

Figure 9-26

7. The Volume of a Pyramid

It can easily be shown that any horizontal cross section of a pyramid is similar to the base. In Figure 9-27, for example, we have a pyramid whose base is the triangular region *abc* and whose cross section is the triangular region *a'b'c'*.

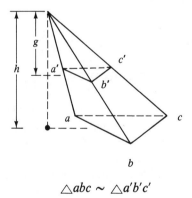

$$\triangle abc \sim \triangle a'b'c'$$

Figure 9-27

If *h* is the height of the pyramid and *g* is the distance from the vertex to the cross-sectional plane, then the constant of proportionality is *g/h*.

$$\frac{a'b'}{ab} = \frac{a'c'}{ac} = \frac{b'c'}{bc} = \frac{g}{h}$$

(See Exercise 9.7, problem 16.)

Recall that if two polygons are similar, with constant of proportionality k, then the ratio of their areas is k^2 (Chapter 8, Section 5). Thus,

$$\frac{\text{Area } (\triangle a'b'c')}{\text{Area } (\triangle abc)} = \left(\frac{g}{h}\right)^2$$

If we have two pyramids with the same height h and the same base area, then we can conclude that horizontal cross sections taken at the same distance g from the vertex will have the same area:

$$\frac{\text{Area } (K)}{\text{Area } (B)} = \left(\frac{g}{h}\right)^2 = \frac{\text{Area } (K')}{\text{Area } (B')}$$

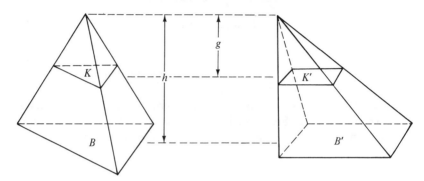

$$\frac{\text{Area } (K)}{\text{Area } (B)} = \left(\frac{g}{h}\right)^2 = \frac{\text{Area } (K')}{\text{Area } (B')}$$

Figure 9-28

Since Area (B) = Area (B'), then Area (K) = Area (K'). By Cavalieri's Principle these two pyramids have the same volume. We thus have the general rule that *two pyramids with the same base area and the same height have the same volume.*

To derive a formula for the volume of a pyramid, let us start with a right triangular prism *abcdef* (Figure 9-29). Since the lateral sides are rectangles, the diagonals divide these sides into pairs of congruent triangles.

Now consider the prism "cut apart" into three pyramids P_1, P_2, and P_3. Take the base of P_2 to be the triangular region *ebf* and the base of P_3 to be region *bcf*. These two bases are congruent, hence they have the same area. The height of both of these pyramids is the distance from their common vertex *d* to the plane containing *bcfe*, hence

$$\text{Vol } (P_2) = \text{Vol } (P_3)$$

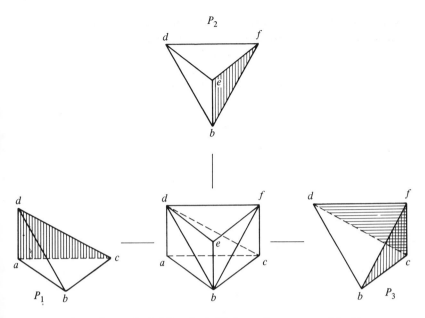

The prism *abcdef* "exploded" into three pyramids P_1, P_2, and P_3

Figure 9-29

Now consider the prism P_1 to have as its base the triangular region *acd*, and this time take the base of P_3 to be the region *dcf*. These two bases also have the same area since they are congruent. The height of these two pyramids is the distance from their common vertex *b* to the plane containing *acfd*, hence

$$\text{Vol}(P_1) = \text{Vol}(P_3)$$

Thus $$\text{Vol}(P_1) = \text{Vol}(P_2) = \text{Vol}(P_3)$$

By the additive property of volumes,

$$\text{Vol}(P_1) + \text{Vol}(P_2) + \text{Vol}(P_3) = \text{Vol}(abcdef) = \text{Area (base)} \cdot \text{height}$$

Since these three volumes are all the same, we have

$$3\text{Vol}(P_1) = \text{Area (base)} \cdot \text{height}$$

or $$\text{Vol}(P_1) = \tfrac{1}{3}\text{Area (base)} \cdot \text{height}$$

We have succeeded in finding a formula for the volume of a special

pyramid—one whose base is triangular and one of whose edges is perpendicular to the base. We can now use our general rule that two pyramids
with the same base area and the same height have the same volume to find
a formula for the volume of *any* pyramid.

Let *P* be a pyramid with base *B* and height *h*.

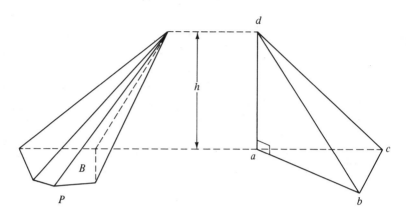

$$\text{Vol } (P) = \tfrac{1}{3} \text{ Area } (B) \cdot h$$

Figure 9-30

We construct a pyramid with triangular base *abc* whose area is equal
to the area of *B*, and with a vertical edge $\overline{ad}$ whose length is equal to *h*.
These two pyramids have the same volume since they have the same base
area and the same height. We have found that

$$\text{Vol } (abcd) = \tfrac{1}{3} \text{ Area } (B) \cdot h$$

therefore $\text{Vol } (P) = \tfrac{1}{3} \text{ Area } (B) \cdot h$

This is a formula for the volume of *any* pyramid *P* whose base is *B*
and whose height is *h*.

Exercise 9.7

1. A half-inch cube is a cube measuring $\tfrac{1}{2}''$ on a side. A half cubic
    inch is half the volume of a cube measuring $1''$ on a side. Is the volume
    of a half-inch cube a half cubic-inch? Explain. Sketch.

2. A solid box has dimensions 2, 3, and 4 feet. What is its volume?
    What is its surface area?

3. A gallon is a measure of liquid volume. It is equivalent to 231 cubic inches. Measure the base and the height of the liquid in a gallon waxed paper carton of milk and compute its volume in cubic inches.

4. What is the smallest number of cuts that would divide a cube of wood whose edge measures 3 inches into cubes each of whose edges measure one inch?

5. If we double each edge of a cube, what effect does this have on its volume? What effect does it have on its surface area?

6. Use the figure below to give a geometric interpretation of the equation

$$(a + b)^3 = a^3 + 3a^2b + 3ab^2 + b^3$$

What is the volume of each numbered piece? An eighth piece is invisible. What is its volume?

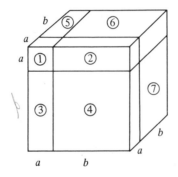

7. Find the volume of a right prism whose base is an isosceles triangle with base 8″ long and height 8″. The height of the prism is 2″. What is its surface area?

8. A feeding trough is 4′ long. Its cross section is an isosceles right triangle with legs 8″ long. Find the volume of the trough in cubic inches; in cubic feet.

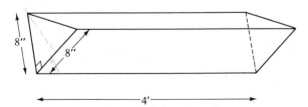

9. The Pyramid of Cheops in Egypt has a square base measuring 720′ on a side. Its height is 480′. What is its volume?

10. A pyramid has a triangular base measuring 3 inches on each side. Its height is 7 inches. What is its volume?

11. A pyramid whose base is a square has volume 27 cubic units. If its height is 9 units, what is the measure of a side of its base?

12. What is the effect on the volume of a pyramid whose base is a square if the height is doubled and each side of the square base is cut in half? If the height is cut in half and each side of the base is doubled?

13. Let $abcd$ be a regular tetrahedron with each edge having length 1 unit. Let $\overline{de}$ be perpendicular to the base and $\overline{bf} \perp \overline{ac}$. Then it can be shown that $eb = \frac{2}{3}fb$.

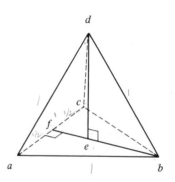

(a) What is the length of $\overline{fb}$?
(b) What is the length of $\overline{eb}$?
(c) What is the length of $\overline{de}$? (*Hint:* $\triangle deb$ is a right triangle.)
(d) What is the area of the base?
(e) What is the volume of the tetrahedron?

14. Suppose that each edge of the regular tetrahedron of problem 13 has length s. Follow the steps outlined to arrive at a formula for its volume.

15. If two pyramids have equal base areas and equal heights, then they have equal volumes. Give the converse of this statement. Is the converse true or false? Give an example to illustrate your answer.

16. Let $abcv$ be a solid pyramid with base in the plane Π_1. Let Π_2 be a plane parallel to Π_1, and let $a'b'c'$ be the cross section of the pyramid in Π_2. Then $\overline{a'b'} \parallel \overline{ab}$; $\overline{b'c'} \parallel \overline{bc}$; $\overline{a'c'} \parallel \overline{ac}$ and $\overline{d'a'} \parallel \overline{da}$.

(a) Show that $\triangle va'd' \sim \triangle vad$.
(b) Show that $\triangle va'b' \sim \triangle vab$.
(c) Show that $va'/va = g/h = a'b'/ab$.

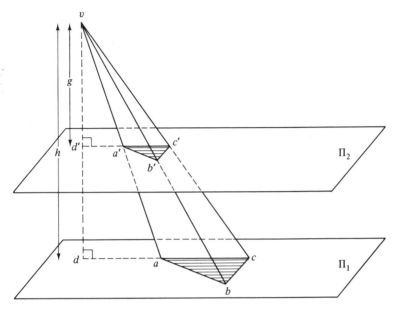

(d) Show that $\triangle vb'c' \sim \triangle vbc$ and $\triangle va'c' \sim \triangle vac$.

(e) Show that $b'c'/bc = a'c'/ac = g/h$.

(f) Conclude that $\triangle abc \sim \triangle a'b'c'$ with constant of proportionality equal to g/h.

17.

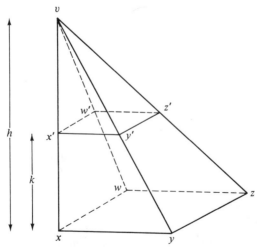

A *truncated pyramid* is a pyramid whose top has been cut off. Consider the truncated pyramid $xyzwx'y'z'w'$ whose lower base $xyzw$ is a square having a side of measure a, and whose upper base $x'y'z'w'$ is a square having a side of measure b. Assume that the upper and lower bases lie in parallel planes. Let h be the altitude of the pyramid and k the distance between the two bases.

(a)   Show that $(h - k)/h = b/a$ (see problem 16) and therefore $h = k(a/(a - b))$.

(b)   Find the volume of the pyramid $vxyzw$.

(c)   Find the volume of the pyramid $vx'y'z'w'$.

(d)   Express the volume of the truncated pyramid as the difference of these two volumes.

(e)   Use (a) to show that this volume is given by the formula

$$V = \tfrac{1}{3} k(a^2 + ab + b^2)$$

18.

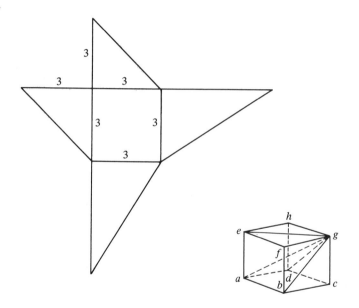

Use the pattern above to construct three pyramids with square bases. Put them together as indicated to form a cube. The three pyramids all have vertex g. Their bases are $abcd$, $abfe$, and $adhe$. If an edge of the cube measures 3 units, what is the volume of each pyramid?

19.   Use the pattern to construct three pyramids P_1, P_2, and P_3. Combine these three pyramids to form a triangular prism as indicated. If the measurements are as given, find the volume of each. What is the volume of the prism?

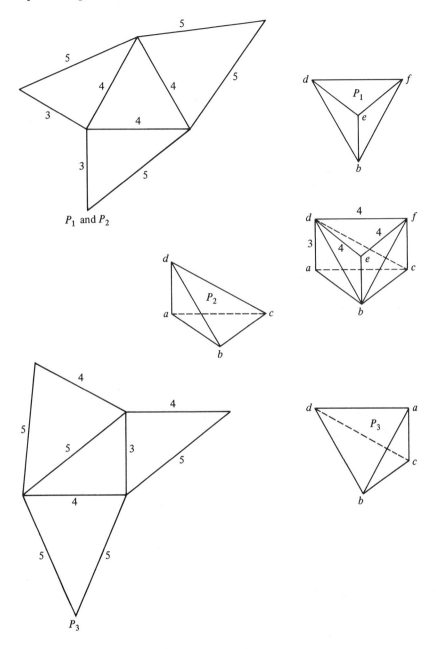

P_1 and P_2

20. State the converse of Cavalieri's Principle. Is the converse true or false? Give an example to illustrate your answer.

21. Let P be a pyramid with base area a and height h. Let B_k be the horizontal cross section at a height k above the base. Show that

$$\text{Area } (B_k) = ((h - k)/h)^2 a$$

    (*Hint:*   See problem 16 and Chapter 8, Section 5.)

22. Find the area of a horizontal cross section of the pyramid of problem 10 three inches above the base.

23. Let P be a pyramid with base area a and height h. Find the area of the mid-section, M. (The mid-section is the horizontal cross section at height $h/2$.)

24. A *prismatoid* is a polyhedron all of whose vertices lie in two parallel planes. The prism and the pyramid are special cases of the prismatoid. The two faces in these parallel planes are called the upper and lower bases, and the cross section parallel to the bases and midway between them is called the mid-section of the prismatoid. Let U, L, and M stand for the areas of the upper base, lower base, and mid-section, respectively, and let h be the distance between the parallel planes. The volume of the prismatoid is then given by the formula

$$V = (h/6)(U + L + 4M)$$

    Show that for the prism and the pyramid this formula reduces to the formulas for their volume we have developed in this chapter. (In the case of the pyramid, $U = 0$.)

25. If two prisms have the same base area and the same height, then their volumes are equal. Will they always have the same surface area? Explain.

New Terms Found in This Chapter

Term	Section	Term	Section
polyhedron	1	lateral faces of a prism	2
faces of a polyhedron	1	parallelepiped	2
edges of a polyhedron	1	altitude or height of a	
vertices of a polyhedron	1	prism	2
regular polyhedron	1	right prism	2
regular tetrahedron	1	rectangular parallelepiped	2
regular hexahedron	1	box	2
regular octahedron	1	cylinder	2
regular dodecahedron	1	circular cylinder	2
regular icosahedron	1	right circular cylinder	2
pyramid	2	sphere	2
altitude or height of a		radius of a sphere	2
pyramid	2	diameter of a sphere	2
lateral faces of a pyramid	2	ball	2
horizontal cross section	2	great circle	2
cone	2	surface area of a	
circular cone	2	polyhedron	Exercise 9.2
right circular cone	2	cubic units	4
slant height	2	volume of a polyhedron	4
prism	2	Cavalieri's Principle	5
lower and upper bases		truncated pyramid	Exercise 9.7
of a prism	2		

10: Measurements Related to Circles

1. The Circumference of a Circle

Before beginning our discussion of measurements related to circles, let us recall the definitions of some terms.

A *circle* is the set of all points in a plane whose distance from some fixed point p (the *center* of the circle) is a fixed positive number r.

A *radius* of a circle is a line segment one of whose end-points is the center of the circle and the other a point on the circle. The measure of such a line segment is also called the radius of the circle.

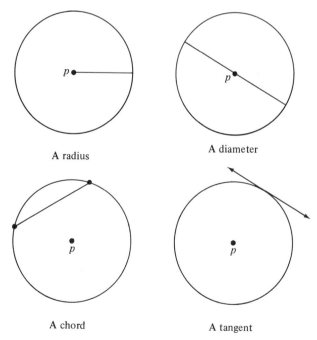

A radius

A diameter

A chord

A tangent

Figure 10-1

A *diameter* of a circle is a line segment passing through the center of the circle whose end-points are on the circle. The measure of this line segment is also called the diameter of the circle. If d stands for the measure of a diameter and r stands for the measure of a radius, then $d = 2r$.

A *chord* of a circle is a line segment whose end-points lie on the circle. A diameter is a special case of a chord.

A *tangent* to a circle is a line which intersects the circle in exactly one point.

An angle whose vertex is at the center of the circle is called a *central angle* of that circle.

An *arc* of a circle is the set consisting of two points on a circle and all the points of the circle between them.

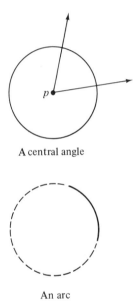

A central angle

An arc

Figure 10-2

It is not easy to define the *circumference* of a circle, although we may have a pretty good idea of what the term means. Intuitively, it is the distance around the circle. If an ant were to walk along the circle starting at a point p, never retracing its steps, until it returned to the point p, the distance traveled would be the circumference of the circle. (Figure 10-3)

If it is difficult to define "circumference," it is even harder to measure it. The circumference is measured in *linear* units such as inches, feet and so on. However, a circle is not straight. It curves constantly. How are we going to measure it with a straight ruler?

The distance the ant travels is the circumference of the circle.

Figure 10-3

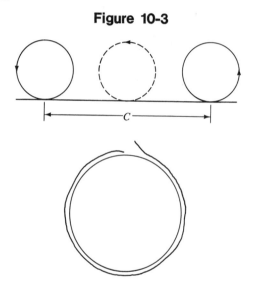

Measuring the circumference of a circle experimentally

Figure 10-4

To measure the circumference of a wheel, we might put a chalk mark on the rim, roll it along a flat surface and then measure the distance on the line between chalk marks.

To measure the circumference of a circle drawn on paper, we might arrange a piece of string carefully so that it lies along the circle, cut it off so that the ends meet, then remove it and lay it on a ruler. These methods will only give us approximations to the measure we want and are not entirely satisfactory. The wheel might slip while we are rolling it along; the string may stretch. What we need is a formula for finding the circumference in terms of some more easily measured length, such as the radius or the diameter of the circle.

You probably learned a long time ago that the circumference of a circle whose radius is r is given by the formula $C = 2\pi r$, (or $C = \pi d$, where d stands for the diameter of the circle) and that the symbol "π" stands for a constant that can be taken to be (approximately) 22/7 or 3.1416.

A rigorous derivation of this formula is given in more advanced mathematics courses. In this section we will try to make this formula seem reasonable and understandable in view of our intuitive understanding of circumference.

Let us start with a regular hexagon whose vertices lie on a circle and whose sides are chords of the circle (Figure 10-5). This is called an *inscribed polygon*.

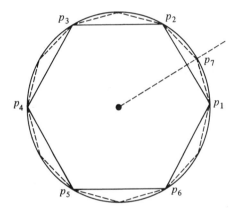

An inscribed regular polygon of six sides; of twelve sides.

Figure 10-5

Clearly, the perimeter of the hexagon is less than the circumference of the circle, C. Call the perimeter of the hexagon P_6. We can draw an inscribed polygon with twelve sides in this same circle by constructing the perpendicular bisector of each side of the hexagon. By the triangle inequality,

$$p_1 p_7 + p_7 p_2 > p_1 p_2$$

thus the perimeter of the polygon with twelve sides, P_{12}, is larger than P_6. P_{12} is, however, smaller than C. We can continue to divide each side into two to obtain P_{24}, P_{48}, P_{96} and so on. Each perimeter is larger than the preceding one, yet they are all smaller than C, the circumference of the circle.

$$P_6 < P_{12} < P_{24} < P_{48} < \cdots < C$$

If it seems strange that you could have a sequence of numbers which are always increasing, yet never get larger than some fixed number, consider the sequence

$$\frac{1}{2}, \frac{2}{3}, \frac{3}{4}, \frac{4}{5}, \frac{5}{6}, \ldots, \frac{n}{(n+1)}, \ldots$$

These numbers get larger and larger, yet they never get any larger than 1.

$$\frac{1}{2} < \frac{2}{3} < \frac{3}{4} < \frac{4}{5} < \frac{5}{6} < \cdots < \frac{n}{(n+1)} < \cdots < 1$$

To go back to the inscribed polygons, it should be clear if you carried out a few of these constructions, that the perimeter of, say a 48-sided polygon, would give a very good approximation to the circumference of the circle. The perimeter of a 96-sided polygon would be even closer, and so on. The sequence of numbers $P_6, P_{12}, P_{24}, P_{48}, \ldots$ yields better and better approximations to C (although no number in this sequence is exactly *equal to* C). We say that *the limit of this sequence is C.*

This deceptively simple phrase involves some rather deep mathematical principles, and we won't attempt to define a limit here. Intuitively, it means that as you proceed from term to term in the sequence, the numbers get closer and closer to C, until they are as close to C as you please.

The sequence

$$\frac{1}{2}, \frac{2}{3}, \frac{3}{4}, \frac{4}{5}, \ldots, \frac{n}{(n+1)}, \ldots$$

has as its limit the number 1, since if we go out along the sequence far enough, we can get a number as close to 1 as we wish. Do we want a term closer to 1 than .1? The 99th term is 99/100 and $1 - 99/100 = .01$. Do we want a term closer to 1 than .01? The 999th term is 999/1000 and $1 - 999/1000 = .001$, and so on.

The circumference of a circle could be *defined* as the limit of the sequence of perimeters of inscribed regular polygons (provided one has defined a *limit*, which we have not done). This is the way the circumference is defined in more advanced geometry courses.

An alternate approach would be to take a sequence of perimeters of *circumscribed* regular polygons. A circumscribed polygon is a polygon each of whose sides is a portion of a tangent to the circle. (See Figure 10-6.)

Let Q_6 stand for the perimeter of the circumscribed regular hexagon. Clearly $Q_6 > C$. If we take a circumscribed regular polygon with twelve sides, its perimeter, Q_{12}, will be smaller than Q_6, but will still be larger

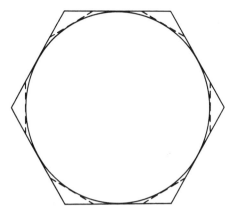

A circumscribed regular polygon of six sides; of twelve sides.

Figure 10-6

than C. In this case the sequence of perimeters is getting smaller, but they are all *larger than* C.

$$Q_6 > Q_{12} > Q_{24} > \cdots > C$$

This is a sequence different from the sequence of inscribed perimeters, but its limit is also the circumference C.

If the formula $C = 2\pi r$ is rewritten as $C/r = 2\pi$, it can be interpreted as saying that the ratio of the circumference of any circle to its radius is a constant. While we are not in a position to verify this statement, we *can* show that *for any circle, the ratio of the perimeter of an inscribed polygon having n sides to the radius is a constant.* Since this is the case it seems reasonable that a similar statement holds for the circumference.

Consider, for example, a regular hexagon inscribed in a circle of radius r.

If we construct the radii to each vertex of the hexagon we will have six congruent isosceles triangles. Let us call the central angle θ. Since these central angles are all congruent and there are six of them, each one has measure $360/6 = 60$. Since the two base angles are congruent, they must each have measure 60, and the triangle is equilateral. Each side of the hexagon then has length r, and the perimeter must be $6r$. The ratio of the perimeter to the radius is $6r/r = 6$, a constant, for any circle whatever. (Figure 10-7)

Now let us consider another case, say an inscribed polygon with ten sides. Again the polygon can be divided into ten congruent isosceles triangles by drawing the radii to the vertices. The central angle, θ, will have measure $360/10 = 36$.

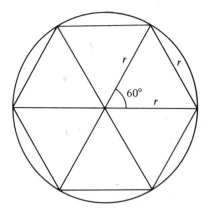

The perimeter of an inscribed regular hexagon $= 6r$

Figure 10-7

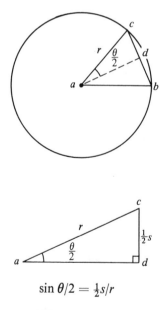

$\sin \theta/2 = \tfrac{1}{2}s/r$

Figure 10-8

Now let us take one of these triangles, *abc*, and divide it into two congruent right triangles by drawing the altitude from the center *a*. Then $\angle cad$ has measure $\theta/2 = 18$, and $cd = \tfrac{1}{2}s$, where *s* is the measure of a side of the polygon. Since adc is a right triangle

$$\sin \frac{\theta}{2} = \frac{\frac{1}{2}s}{r}$$

(Figure 10-8), thus $s = 2r \sin(\theta/2) = 2r \sin 18°$, and the perimeter is $10s = 20r \sin 18°$. The ratio of the perimeter to the radius is $20 \sin 18° \doteq 6.18$. (Compare this to $2\pi \doteq 6.283$.)

In general, if the inscribed polygon has n sides, the central angle θ will be $360/n$; s will be $2r \sin(360/2n)$, and the perimeter will be $n[2r \sin(360/2n)]$. The ratio of this perimeter to the radius is a constant, $2n \sin(360/2n)$.

Exercise 10.1

In the following problems if you are asked for approximate distances, use $\pi \doteq 3.1416$.

1. Find the approximate circumference of a circle with radius 2; with diameter 5. Find the approximate radius of a circle if its circumference is 25.

2. The earth travels in an orbit which is nearly circular with the sun at the center. If the radius of the path is 93,000,000 miles, approximately, how far does the earth travel in one orbit? If the earth makes one orbit in 365 days, what is its approximate speed in miles per hour?

3. If the radius of a circle is doubled how does this affect the diameter? How does it affect the circumference?

4. Let n be the number of sides of a regular polygon inscribed in a circle of radius r. Let s be the measure of each side ($s = 2r \sin \theta/2$). Divide the polygon into congruent triangles by constructing the radii to its vertices and let θ be the measure of each of these central angles. Use the table of trigonometric ratios on p. 162 to complete the following table:

n	central angle (θ)	$\theta/2$	$\sin(\theta/2)$	s	Perimeter $= ns$	Ratio of perimeter to radius
3	$\frac{360}{3} = 120$	60	.8660	1.732r	5.196r	5.196
4						
5						
6						
12						

5. Compare the values in the last column of the table of problem 4 to $2\pi \doteq 6.2832$. What kind of value do you think you would get for the ratio if you took n to be very large?

6. Let $p_1 p_2 p_3 \cdots p_n$ be a regular polygon of n sides inscribed in a circle with center o and radius r. Prove that $\triangle p_1 o p_2 \cong \triangle p_2 o p_3$. What is the measure of the central angle θ?

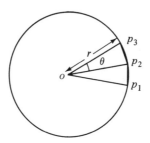

7. Find the radius of a circle if an inscribed square has a side 2 inches long. What is its circumference?

8. The circumference of a circle is 15 inches. If this is increased to 18 inches, approximately how much does the radius change?

9. If the radius of a circle is increased by two units, how much does the circumference change? (*Hint:* Let C_1 and r_1 stand for the original circumference and radius and let C_2 be the new circumference. Then, $C_1 = 2\pi r_1$, and $C_2 = 2\pi(r_1 + 2) = 2\pi r_1 + 4\pi$.)

10. Suppose that a steel band is stretched tightly around the earth at the equator so that it touches everywhere. (Assume the earth is level at the equator.) Approximately how much would the length of the band have to be increased so that a six-foot-tall man could walk under it at any point on the equator?

11. Explain why the answer to problem 10 would be exactly the same if instead of the earth, the band were stretched around a sphere of any size.

2.   The Number π

The ratio of the circumference of any circle to its diameter is a constant. We use the Greek letter π for this constant. In ancient times this number was taken to be 3.

Probably the first mathematical attempt to find this constant was made by Archimedes about 240 B.C. He knew that the perimeter of any *inscribed* regular polygon was *less than* the circumference of the circle, while the perimeter of a *circumscribed* regular polygon was always *greater than* the circumference.

In a circle whose radius is 1, for example, the perimeter of an inscribed regular hexagon is 6, while the perimeter of a circumscribed regular hexagon is $12/\sqrt{3} = 4\sqrt{3} \doteq 6.928$. The circumference of the circle, which is 2π, is between these two numbers.

$$6 < 2\pi < 6.928$$

or
$$3 < \pi < 3.464$$

We might use as an approximation for π the average of these two numbers, or 3.232. To get a better approximation we could use inscribed and circumscribed regular polygons of 12, 24, 48 or more sides. This is what Archimedes did, and he found that π was between 223/71 and 22/7, or 3.14 to two decimal places. This method of computing π is called the *classical method.* Following Archimedes many other mathematicians used the classical method to find a more accurate value for π. In 1610 Ludolph van Ceulen of Germany computed π to 35 decimal places using polygons having 2^{62} sides. He spent most of his life on this project and when he died the number, called the Ludolphian number, was engraved on his tombstone.

Late in the 17th century, mathematicians took a different approach to the problem. The infinite series

$$\frac{\pi}{4} = 1 - \frac{1}{3} + \frac{1}{5} - \frac{1}{7} + \cdots$$

called *Gregory's series,* after the Scottish mathematician James Gregory, was found, and with various modifications has been used ever since to find approximations to π.

If you add the first 50 terms of Gregory's series, for example,

$$1 - \tfrac{1}{3} + \tfrac{1}{5} - \tfrac{1}{7} + \cdots - \tfrac{1}{99}$$

you get a value for $\pi/4$ with an error less than 1/101. Using a modification of this method in 1873, William Shanks of England computed π to 707 places. This monumental achievement took him 15 years.

With the advent of the computer, computations which had previously taken years to do were made in minutes. The first computer to be used on the problem of computing π was the ENIAC, the Electronic Numerical Integrator and Calculator, in 1949. Using the ENIAC, π was calculated to 2,037 decimal places in 70 hours. Newer computers have been able to perform the calculations much more rapidly.

Machine	Date	Decimal Places	Time
ENIAC	1949	2,037	70 hours
IBM 704	1958	10,000	100 minutes
IBM 7090	1961	100,265	8.7 hours

Why did so many mathematicians spend so much time computing π to more and more decimal places? A ten decimal place value for π gives the circumference of the earth within a fraction of an inch. Thirty decimal places would give the circumference of the entire observable universe with an error too small to be measured by the most powerful telescope. In applications, the most careful work requires only 4 decimal places. Why then, compute π to 100,000 decimal places?

Early mathematicians were looking for a repeating pattern in the decimal. If such a pattern had been found, this would have meant that π was a rational number and could be expressed exactly as the ratio of two integers. No such pattern was found, and in 1767 Lambert showed that π was an irrational number, hence its decimal can never repeat. Since the invention of the electronic computer, each new computer has been assigned the task of calculating π as a test to demonstrate its speed and accuracy. It has been predicted that future generations of computers will be many times as fast as those we have now.

Strangely enough, the number π arises in other ways than as the ratio of the circumference of a circle to its diameter. The famous Buffon Needle Problem demonstrates that π is found in probability theory. If a number of parallel lines are drawn 2 inches apart on a flat surface, and a needle 1 inch long is dropped at random on this surface, then the probability that the needle will fall across one of the lines is $1/\pi$. Some mathematicians have actually computed π experimentally in this way.

No discussion of the number π would be complete without the story of the attempt in 1897 by the Indiana State Legislature to pass a bill setting the value of π at 4. The attempt failed although it was supported by the State Superintendent of Public Instruction.

Exercise 10.2

1. The following poem by A. C. Orr is actually a device for remembering π to 30 decimal places. Find this value of π by replacing each word by the number of letters it contains.

> Now I, even I, would celebrate
> In rhymes unapt, the great
> Immortal Syracusan, rivaled nevermore,

> Who in his wondrous lore,
> Passed on before,
> Left men his guidance
> How to circles mensurate.*

2. The number 22/7 is often used as an approximation to π. How accurate is this? Compare to the value of π obtained in problem 1. In about the year 480, the Chinese used the value 355/113 for π. How accurate is this? In the Middle Ages $\sqrt{10}$ was frequently used as an approximation to π. How accurate is this?

3. House Bill No. 246, Indiana State Legislature, 1897, states (incorrectly) that if a circle and a square have equal perimeters, then they have equal areas. Show that if this is true then $\pi = 4$. (*Hint:* If the perimeter of a square is $2\pi r$, the measure of one of its sides is $(2\pi r)/4 = \pi r/2$, and its area is $(\pi r/2)^2$.)

4. In I Kings 7: 23 we read, "And he made a molten sea, ten cubits from the one brim to the other: . . . and a line of thirty cubits did compass it round about." What value for π do these measurements imply?

3. The Area of a Circle

Long ago you learned that the area of a circular region was given by the formula $A = \pi r^2$, where r is the radius of the circle. This formula is derived in more advanced mathematics courses. We will not do so. However, in this section we will try to show that this formula is a reasonable one.

Suppose that we cut a circle into a number of congruent pieces, each shaped like a piece of pie (these pieces are called *sectors*) and arrange them as shown in Figure 10-9.

This figure looks a little like a parallelogram—its area should be close to that of a parallelogram having base b and altitude h. Now the height of this "parallelogram" is equal to r, the radius of the circle, and the base b is approximately half the circumference, or $(2\pi r)/2 = \pi r$. The area of this figure then should be close to $b \cdot h = r(\pi r) = \pi r^2$.

If we had used 20 sectors instead of 10, the figure would have looked even more like a parallelogram, and b would be even closer to πr.

We can approximate the area of a circle by finding the area of an inscribed regular polygon.

We saw in Chapter 8, Section 3 that the area of a regular polygon is given by the formula Area $= \frac{1}{2}a \cdot P$, where P is the perimeter of the polygon

---

Literary Digest, 1906. Copyright Time, Inc.

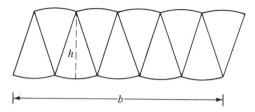

Ten congruent circular sectors rearranged to form a "parallelogram"

Figure 10-9

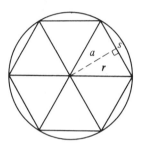

Area of a regular hexagon. $6(\frac{1}{2}a \cdot s) = \frac{1}{2}a (6s) = \frac{1}{2}aP$.

Figure 10-10

and a is the apothem, the perpendicular distance from the center to one of the sides.

If we increase the number of sides, the area of the inscribed polygon will be greater, but will still be less than the area of the circle. If A_n stands for the area of an inscribed polygon of n sides, we have the series of inequalities

$$A_3 < A_4 < A_5 < \cdots < A_n < \cdots < A$$

where A is the area of the circle. This is a sequence of numbers which gets closer and closer to the area of the circle. We say that the *limit* of this sequence is A.

Now if P_n stands for the perimeter of the inscribed polygon with n sides, and a_n stands for its apothem, then

$$A_n = \frac{1}{2}a_n P_n$$

and we have

$$\frac{1}{2}a_3 P_3 < \frac{1}{2}a_4 P_4 < \frac{1}{2}a_5 P_5 < \cdots < \frac{1}{2}a_n P_n < \cdots < A$$

We saw in Section 1 that the sequence $P_3, P_4, P_5, \ldots$ of perimeters approaches the circumference of the circle, $2\pi r$, and it is not hard to see that the sequence $a_3, a_4, a_5, \ldots$ of apothems gets closer and closer to the radius of the circle. The limit then of the sequence $\frac{1}{2}a_3P_3, \frac{1}{2}a_4P_4, \ldots$ is $\frac{1}{2}(r)(2\pi r) = \pi r^2$, the area of the circle.

A *sector* of a circle is a region bounded by two radii and an arc of the circle.

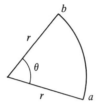

A sector of a circle with central angle θ

Figure 10-11

The area of a sector is closely related to the measure of the central angle θ. Consider the sectors in Figure 10-12.

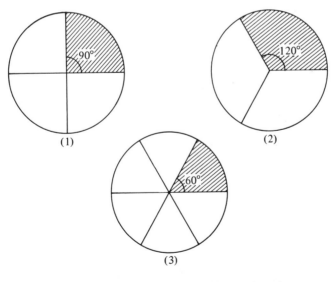

$$\frac{\text{Area of sector}}{\text{Area of circle}} = \frac{\text{Measure of central angle}}{360°}$$

Figure 10-12

In (1) the area of the sector is $\frac{1}{4}$ the area of the circle and its central angle is 90°, or $\frac{1}{4}(360°)$. In (2) the area is $\frac{1}{3}$ the area of the circle and the central angle measures $120° = \frac{1}{3}(360°)$. This suggests the following rule:

The ratio of the area of a circular sector to the area of the circle is equal to the ratio of the measure of its central angle to 360°.

$$\frac{\text{Area of sector}}{\text{Area of circle}} = \frac{\text{Measure of central angle}}{360°}$$

or $\text{Area of sector} = \dfrac{\text{Measure of central angle}}{360°} \cdot \text{Area of circle}$

Thus, if a sector of a circle of radius 3 has a central angle of 60°, then the area of the sector is

$$\tfrac{60}{360} \cdot \text{Area of the circle} = \tfrac{1}{6}(\pi 3^2) = 3\pi/2 \doteq 4.7$$

Exercise 10.3

1. If the radius of a circle is $2\frac{1}{2}$, find its area. If the area of a circle is 2π square inches, what is its radius?

2. If d is the diameter of a circle, show that its area is given by the formula $A = \pi d^2/4$.

3. The circumference of a circle is 22 inches. What is its area? (Use $\pi \doteq 22/7$.)

4. The area of a circle is 154 square inches. What is its circumference? (Use $\pi \doteq 22/7$.)

5. The area of a circle C is equal to the sum of the areas of circles C_1 and C_2. If C_1 has radius 3 and C_2 has radius 4, what is the radius of circle C?

6. A circle is inscribed in a square whose side measures 2 inches. What is the radius of the circle? What is its area?

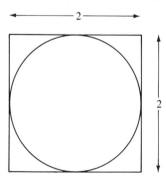

7. A square is inscribed in a circle whose radius is 2 inches. What is the measure of one side of the square? What is its area?

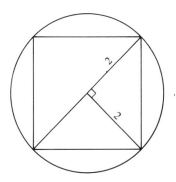

8. A square whose side is 2 inches is inscribed in a circle. What is the radius of the circle? What is its area?

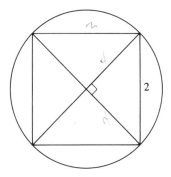

9. Find the area of the shaded region.

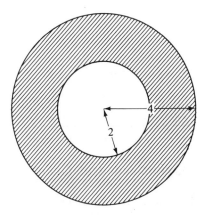

10. Find the area of the shaded region.

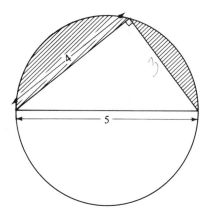

11. Find the area of the shaded region. Find the area of the unshaded region.

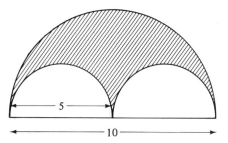

12. Find the area of the shaded region.

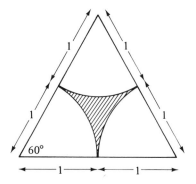

4. Cylinders and Cones

If two geometric solids have their bases in the same plane Σ, and if every plane parallel to Σ cuts these solids in cross sections which have the same area, then by Cavalieri's Principle (Chapter 9, Section 5) the two solids have the same volume. This principle holds no matter what kind of region the cross sections may be. For example, the cross sections of one solid might be rectangular regions, and those of the other might be circular regions.

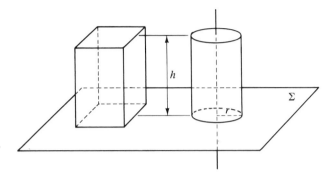

Volume of the cylinder $=$ Volume of the box
$$= \text{Area of base} \cdot h$$
$$= \pi r^2 h$$

Figure 10-13

The cross sections of a *right circular cylinder* parallel to the base are all congruent circles. If we have a solid box whose base has the same area as the circle, πr^2, and whose height h is the same as the height of the cylinder, then

Volume of the cylinder $=$ Volume of the box
$$= \text{Area of the base} \cdot \text{height}$$
$$= \pi r^2 h$$

We have then a formula for the volume of a right circular cylinder

$$\text{Volume} = \pi r^2 h$$

where r is the radius of the base and h is the height. It follows from Cavalieri's Principle that this formula gives us the volume of *any* circular cylinder, whether it is a *right* circular cylinder or not.

If we want to fill a tank in the shape of a right circular cylinder, we are interested in knowing its volume, but if we want to paint the outside of the tank, we need to know the surface area we have to cover.

The total surface area of a right circular cylinder can be thought of as three pieces: top, bottom, and sides. The top and bottom are congruent circles, each having area πr^2. The area of the sides is called the *lateral sur-*

face area. To measure the lateral surface, imagine that you remove the top and bottom of the cylinder, then cut down one side and open it out, much as you might remove the label from a can. (Figure 10-14)

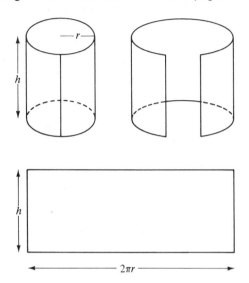

Measuring the lateral surface area of a right circular cylinder

Figure 10-14

The resulting region will be rectangular. Its width is h, the height of the cylinder, and its length is $2\pi r$, the circumference of the base. The lateral surface area is then $(2\pi r)h$, and the *total* surface area (sides, bottom, and top) is given by the formula

$$\text{Surface Area} = 2\pi rh + 2\pi r^2$$

We found in Chapter 9 that the volume of any pyramid was given by the formula

$$\text{Volume} = \tfrac{1}{3} \cdot \text{Area of the base} \cdot \text{height}$$

Given a (solid) right circular cone whose base lies in some plane Σ, a plane parallel to Σ will cut the cone in a circular region. If we have a pyramid, say with a square base whose area is the same as the area of the base of the cone, πr^2, and whose height h is the same as that of the cone, then it can be shown that any plane parallel to Σ will cut these two solids in cross sectional regions which have the same area. By Cavalieri's Principle it follows that these two solids have the same volume and

Volume of the cone $=$ Volume of the pyramid

$= \frac{1}{3} \cdot$ Area of the base $\cdot$ height

$= \frac{1}{3} \cdot \pi r^2 \cdot h$

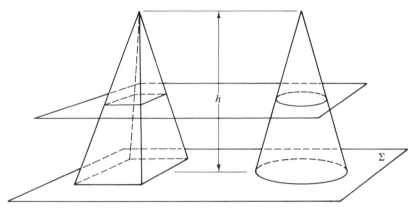

Volume of the cone $=$ Volume of the pyramid

$= \frac{1}{3} \cdot$ Area of base $\cdot$ height

$= \frac{1}{3} \cdot \pi r^2 \cdot h$

Figure 10-15

We now have a formula for the volume of a right circular cone.

$$\text{Volume} = \frac{1}{3} \cdot \pi r^2 \cdot h$$

where r is the radius of the base and h is the height. It follows from Cavalieri's Principle that this formula gives us the volume of *any* circular cone, whether it is a *right* circular cone or not.

Now let us derive a formula for the surface area of a right circular cone. The base, of course, has area πr^2. To find the lateral surface area, picture the cone opened out as in Figure 10-16.

Since the circumference of the base is $2\pi r$, this must be the length of the arc ab. The length l is called the *slant height* of the cone. The lateral surface whose area we are trying to find is the circular sector which is a part of the circular region whose total area is $\pi \cdot l^2$. What part? Well, it seems reasonable that the longer the arc ab the greater the area of the sector. In fact, the ratio of the area of the circular sector to the total area of the circle will be the same as the ratio of the length of the arc ab to the circumference of the circle. Thus, we have

$$\frac{\text{Area of sector}}{\pi l^2} = \frac{2\pi r}{2\pi l}$$

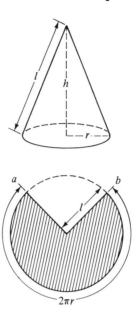

Measuring the lateral surface area of a right circular cone

Figure 10-16

or $$\text{Area of sector} = \pi l^2\left(\frac{r}{l}\right)$$

$$= \pi r l$$

The lateral surface area of a right circular cone is thus given by the formula

$$\text{Area} = \pi r l$$

where r is the radius of the base and l is the slant height of the cone.

Since this is a right circular cone, the altitude is perpendicular to the radius and l is the length of the hypotenuse of a right triangle. (Figure 10-16) The radius r, the height h, and the slant height l are related by the equation

$$r^2 + h^2 = l^2$$

Thus if we know r and h, we can find the slant height l.

For example, suppose we want to find the lateral surface area of a cone for which $r = 3''$ and $h = 4''$. Then $l = \sqrt{3^2 + 4^2} = \sqrt{25} = 5''$, and the lateral surface area is $\pi 3'' \cdot 5'' \doteq 47$ square inches.

Exercise 10.4

1. Complete the following table for right circular cylinders having the measures given.

radius of base	height	volume (cubic units)	lateral surface area (square units)	total surface area (square units)
2	8			
15		450π		
6		216π		
	$\frac{1}{2}$		2π	

2. What is the effect on the volume of a right circular cylinder if the height is doubled? If the radius of the base is doubled? If the radius is doubled and the height cut in half?

3. What is the effect on the lateral surface area of a right circular cylinder if the height is doubled? If the radius of the base is doubled? If the radius is doubled and the height cut in half?

4. An open pipe made of metal is 12 inches long. Its inner radius is 1 inch and it is $\frac{1}{4}$ inch thick. Find the volume of the metal in this piece of pipe.

5. One U.S. gallon of liquid has a volume of 231 cubic inches. If a cylindrical jar has a base whose radius is 2 inches, how tall must it be in order to hold 1 gallon? (Use $\pi \doteq 22/7$.)

6. Complete the following table for the right circular cones having the measures given.

radius	height	volume	slant height	lateral surface area	total surface area
2	9				
8		64π			
	$\frac{1}{2}$	$(\frac{2}{3})\pi$			
3			$3\sqrt{2}$		

7. What is the effect on the volume of a cone if the height is doubled and the radius left unchanged? If the radius is doubled and the height left unchanged? What is the effect on the slant height if the radius and the height are both doubled? What is the effect on the lateral surface area

if the radius and the height are both doubled? What is the effect on the volume if the radius and the height are both doubled?

5. The Sphere

Now let us consider the problem of finding the volume of a sphere. You probably have been told that the formula for this volume is $(\frac{4}{3})\pi r^3$, where r is the radius of the sphere.

First, let us look at the problem intuitively and see that this formula is a reasonable one. Suppose that a sphere of radius r is enclosed in a cube whose edges measure $2r$.

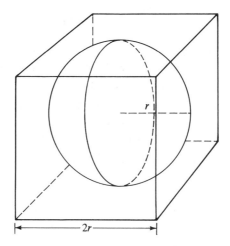

The volume of a sphere of radius r is approximately half the volume of the cube measuring $2r$ on a side.

Figure 10-17

The volume of the sphere is clearly less than that of the cube; we might estimate it to be about half as much. This is actually a fairly accurate estimate, since the volume of the cube is $(2r)^3 = 8r^3$, and half of this is $4r^3$. This is close to $(\frac{4}{3})\pi r^3$.

We can arrive at the formula exactly by using Cavalieri's Principle. To simplify matters, let us find the volume of half a sphere (a hemisphere).

A horizontal cross section through the hemisphere at a distance h above the base will be a circle. The radius of this circle, x, will depend on how far above the base you take the cross section. If you take it very close to

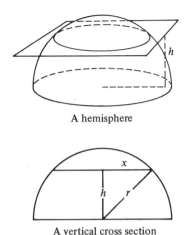

A hemisphere

A vertical cross section

Figure 10-18

the base, x will be close to r. If you take it close to the upper boundary of the hemisphere, it will be very small. If we look at a vertical cross section of the hemisphere we can see that h, r, and x are all related, since they are the sides of a right triangle.

$$h^2 + x^2 = r^2$$

or $$x^2 = r^2 - h^2$$

The area of any cross section of the hemisphere taken at a height h above the base is then $\pi x^2 = \pi(r^2 - h^2)$.

Now let us look at another solid, a right circular cylinder whose radius is r and whose height is also r. From this solid cylinder a right circular cone is removed. The base of the cone is the upper base of the cylinder and its vertex is at the center of the lower base.

The solid, a vertical cross section through the center of the solid, and a horizontal cross section at a height h above the base are shown in Figure 10-19. The horizontal cross section looks like an annulus, or ring. The radius of the *outer* circle is r, and from the vertical cross section we can see that the radius of the *inner* circle is h. The area of this cross section then is $\pi r^2 - \pi h^2 = \pi(r^2 - h^2)$. This is the same as the cross sectional area at height h of the hemisphere. By Cavalieri's Principle the two solids must have the same volume. We can easily compute the volume of the cylinder with the cone removed, since we have formulas for both these volumes. The volume of the cylinder is $\pi r^2 \cdot r = \pi r^3$ (recall that its height was taken to be equal to r) and the volume of the cone is $(\frac{1}{3})\pi r^2 \cdot r = (\frac{1}{3})\pi r^3$.

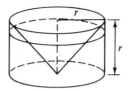

A cylinder with a cone removed

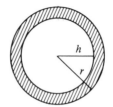

A vertical cross-section

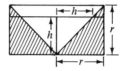

A horizontal cross-section

Figure 10-19

The difference is $\pi r^3 - (\frac{1}{3})\pi r^3 = (\frac{2}{3})\pi r^3$. The volume of the hemisphere then must be $(\frac{2}{3})\pi r^3$. Since this is *half* the volume of the sphere, we arrive at the formula

$$\text{Volume of the sphere} = (\tfrac{4}{3})\pi r^3$$

The surface area of the sphere is given by the formula

$$\text{Area} = 4\pi r^2$$

where r is the radius of the sphere. To derive this formula rigorously, we would need calculus; however, we can see intuitively that it is a reasonable one.

Again consider the sphere enclosed in a cube whose side is $2r$. (Figure 10-17) The surface area of the sphere is less than the surface area of the cube—we might hazard a guess that it is about half as much. The cube has six faces, each of which has area $(2r)^2$. The total surface area of the cube then is $6 \cdot 4r^2 = 24r^2$. Half of this is $12r^2$, and this is very close to $4\pi r^2$.

Exercise 10.5

1. Complete the following table for the spheres having the measures given.

radius	volume	surface area
2		
	36π	
		π
10		

2. What happens to the volume of a balloon shaped like a sphere when its radius is doubled? What is the effect on its surface area?

3. Find the volume of a right circular cone whose radius and height are each 1 inch; a right circular cylinder whose radius and height are each 1 inch; a hemisphere whose radius is 1 inch. What relationship do you observe between these three numbers?

4. What is the total surface area of a hemisphere of radius r? What is its volume?

5. A water tank in the shape of a hemisphere has a radius of 10 feet. How many gallons will it hold? (231 cubic inches = 1 gallon)

6. If a gallon of paint covers 300 square feet how much paint will it take to paint the outside (sides and top) of the hemispherical water tank of problem 5?

7. Find the volume of the largest sphere which can be placed inside a cube whose edge is 1 foot long.

8. If the radius of the earth is 4,000 miles, compute the surface area of the earth. Assume the earth is a sphere and use $\pi \doteq 3.14$.

New Terms Found in This Chapter

Term	Section	Term	Section
circle	1	limit of a sequence	1
radius	1	circumscribed polygon	1
diameter	1	sector of a circle	3
chord	1	lateral surface area of	
tangent to a circle	1	a cylinder	4
central angle of a circle	1	slant height of a cone	4
arc of a circle	1	lateral surface area of a	
circumference of a circle	1	cone	4
inscribed polygon	1	hemisphere	5

11: Coordinate Geometry

1. Introduction

A major contribution to mathematics was made in 1637 when René Descartes published his *Discours*, a philosophical work on universal science. A short appendix to this work was called *La géométrie*. This contains the basis of what we now call analytic geometry. This contribution, which virtually revolutionized geometry, consisted of the bringing together of algebra and geometry in a way which combined the best features of both. Through this marriage of algebra and geometry we can solve geometric problems by algebraic methods, or conversely use geometric methods to throw light on algebraic problems.

Essentially, the basic idea of analytic geometry, or coordinate geometry, is that there is a one-to-one correspondence between ordered pairs of real numbers and points in a plane. The result of this is a correspondence between algebraic sentences such as $x^2 + y^2 = 1$, $x + 2y = 4$, or $0 \leq x \leq 1$, and geometric figures in a plane. In studying these figures we can use both the manipulation of symbols, which is a strong point of algebra, and the figures which serve as an aid to intuition in geometry.

A story akin to the one about Newton and the falling apple is told about Descartes and the discovery of coordinate geometry. It is said that the idea occurred to him as he was lying in bed watching a fly crawl on the ceiling near a corner of his room. He realized that he could describe the path of the fly if he knew some rule relating the fly's distances from the two near-by walls.

2. Coordinates of Points

According to the Ruler Axiom there is a one-to-one correspondence between the real numbers and points on a line. This concept of a number line is extremely useful. Given a point on a line to which we assign the number zero and a unit length, we can locate any point on the line very conveniently by giving the real number assigned to it. This number is called the *coordinate* of the point.

Now suppose we have some point p in the plane. How can we describe its position? To locate points in the plane we turn to the idea of the Cartesian product (named after Descartes) of two sets. (Chapter 1, Section 2) If R stands for the set of real numbers, then $R \times R$ is the set of ordered pairs (x, y) where x and y are elements of R. There is a one-to-one correspondence between points in the plane and elements of $R \times R$. Using this correspondence we can describe precisely the location of any point in the plane.

We start out in Figure 11-1 with a number line which we will call the x-axis. Now through the point labeled zero we draw a second line perpendicular to the first. We call this second line the y-axis. Now we set up a coordinate system on the y-axis with the point of intersection being assigned the number zero. This point is called the *origin*.

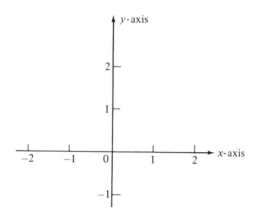

The coordinate axes

Figure 11-1

The usual practice is to assign positive numbers to points on the x-axis to the right of the origin and to points on the y-axis above the origin. We use arrows to indicate the positive directions.

Now we are ready to assign to every point p in the plane a unique ordered pair of real numbers, and we will do this in the following way. Through the point p construct a line perpendicular to the x-axis (Figure 11-2). This perpendicular will intersect the x-axis in a point a. Since a is a point on a number line, there is a real number assigned to it; let's call it x. The number x will be called the x-coordinate or the *abscissa* of the point p.

In the same way, construct a line through p perpendicular to the y-axis, intersecting the y-axis in the point b. Since b is a point on a number line, there is a real number assigned to it, say y. The number y is the y-coordinate or *ordinate* of the point p.

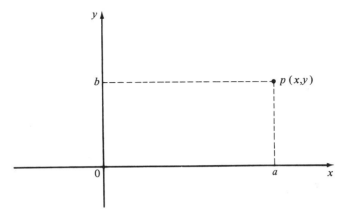

Rectangular coordinates of p

Figure 11-2

We therefore assign to the point p the ordered pair of real numbers (x, y). These real numbers are called the *coordinates of the point p*. It is customary in giving the coordinates of a point to list the x-coordinate first, followed by the y-coordinate. These are also called *rectangular coordinates* because the figure whose vertices are the origin, a, p, and b is a rectangle.

Points on the x-axis will be assigned to number pairs of the type $(x, 0)$ while points on the y-axis will be described by number pairs of the form $(0, y)$. The origin has coordinates $(0, 0)$.

This gives us a plan for assigning to every point in the plane a member (x, y) of $R \times R$. Now, conversely, given any ordered pair of real numbers we can find the point in the plane corresponding to it.

To find the point in the plane corresponding to the ordered pair $(2, 3)$, we construct a perpendicular to the x-axis at the point labeled 2, and a perpendicular to the y-axis at the point labeled 3. The point of intersection of these two perpendiculars is the point we seek. A few examples (see Figure 11-3) will make our correspondence clear.

In view of this one-to-one correspondence, in the future we will refer to "the point (x, y)." Note that these are *ordered* pairs. The point $(2, 3)$ is not the same as the point $(3, 2)$.

Although we have chosen our x and y-axes to be perpendicular lines, this is not essential. In some fields of mathematics curves are used as axes,

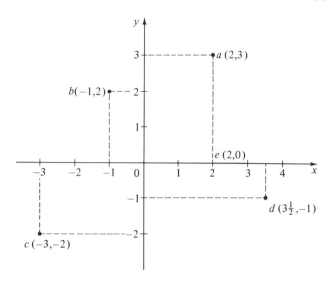

Assigning points in the plane to ordered pairs of numbers

Figure 11-3

and non-perpendicular intersecting lines may also be used. The choice of positive direction is purely arbitrary. We could just as well choose the positive direction to be down or to the left. Finally, the choice of scale on the two axes is left to the discretion of the student. Although it will usually be convenient to use the same unit length on both, sometimes this is not practical. For example, if you want to plot the point (1, 10000) it would be common sense to use different unit lengths on the x- and y-axes. As long as your scale is *clearly indicated* on both axes, this is perfectly acceptable.

In working the problems in this chapter you will find it convenient to use graph paper. Axes may be drawn anywhere on the sheet. Always label your axes x and y and include the arrows to indicate the positive directions. It is not necessary to label every unit on your graph—in fact, this is not particularly desirable. Label the origin and enough points on your axes so that you can read off the coordinates of the points you are plotting.

Exercise 11.2

Use graph paper for these problems.

1. Draw and label a pair of axes as shown in the previous section. Plot the following points:

(a) (1, 3)   (b) (−2, 1)
(c) (−3, −5)   (d) (0, −6)
(e) (0, 0)   (f) (5, 5)
(g) (−5, −5)   (h) (5, −5)
(i) (−5, 5)   (j) (−6, 0)

2. Plot each of the following points on a coordinate system.

(a) (1, 1)   (b) (1, 3)
(c) (1, −4)   (d) (1, −6)
(e) (1, 0)

Where would you say that all points with x-coordinate 1 could
be found?

3. Plot each of the following points on a coordinate system.

(a) (1, −1)   (b) (−1, −1)
(c) (5, −1)   (d) (−6, −1)
(e) (0, −1)

Where you would say that all points with y-coordinate −1 could be
found?

4. Plot the following pairs of points:

(a) (1, 3); (3, 1)
(b) (−2, −4); (−4, −2)
(c) (3, −1); (−1, 3)

Does interchanging the x-coordinate and y-coordinate change the
point? Will interchanging the x-coordinate and y-coordinate always
change the point? Name at least 5 points such that interchanging the
x and y-coordinates does not change the point. Plot them on your
graph.

5. Describe the following sets and plot a few representative elements of
each:

(a) the set of all points with abscissa negative.
(b) the set of all points with ordinate greater than or equal to
zero.
(c) the set of all points with abscissa zero.
(d) the set of all points with ordinate zero.
(e) the set of all points with abscissa and ordinate both zero.
(f) the set of all points with at least one coordinate zero.
(g) the set of all points for which the abscissa is equal to the
ordinate.
(h) the set of all points for which the abscissa is twice the ordinate.

6. Plot the points $(1, 3)$, $(2, 4)$, $(3, 5)$, $(4, 6)$. Note that these points all lie on a line. What is the relationship between the abscissa and the ordinate of each of these points? Can you predict what some other points on this line will be?

7. Answer the same questions as in problem 6 for the points $(-1, -2)$, $(0, 0)$, $(1, 2)$, $(2, 4)$.

8. Plot the points a $(1, 3)$ and b $(2, 2)$. Find the coordinates of a third point c so that abc is a right triangle with right angle at c. (There are two answers to this problem.)

9. Plot the points a $(1, 1)$, b $(1, -2)$ and c $(-3, -2)$. Find the coordinates of a fourth point d so that $abcd$ is a rectangle.

10. Plot the points a $(3, 2)$ and b $(3, -4)$. Find the coordinates of two more points c and d so that $abcd$ is a square. (There are two answers to this question.)

11.

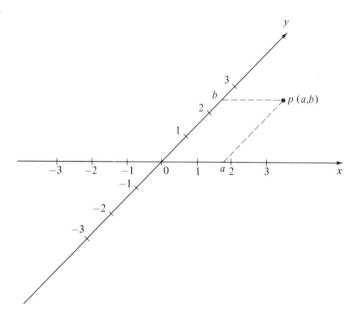

Let us devise a *parallelogram coordinate* system in the following way. Suppose the x-axis and the y-axis intersect at an angle of $45°$. Given any point p in the plane, we assign to it x and y-coordinates by drawing through p lines parallel to the y-axis and to the x-axis. The first intersects the x-axis in a point having coordinate a and the second intersects the y-axis in a point having coordinate b. The *parallelogram coordinates* of p are (a, b). Locate the following points in this system: $(1, 0)$, $(-2, -1)$, $(1, -3)$, $(-2, 2)$.

3. Graphs

Now we are ready to examine further the correspondence between geometric figures and algebraic sentences. Some examples of algebraic sentences involving x and y are $x + y = 1$; $x > y$; $x = 2$. Although this last sentence may not seem to be a sentence about both x and y, we will regard it as saying "$x = 2$ and y can be any real number"; or to put it another way, "$x = 2$ and y is unrestricted."

Corresponding to each geometric figure we study there will be one or more equations or inequalities, and conversely, every equation or inequality or combination of these may describe a geometric figure in the plane. This correspondence is not one-to-one. For example, the two sentences

$$x + y = 0$$

and $$2x + 2y = 0$$

correspond to the same geometric figure (a line).

On the other hand, some algebraic sentences describe no figure at all. Consider, for example, the pair of inequalities

$$x < 2 \quad \textit{and} \quad x > 5$$

These two inequalities are contradictory. There is no real number which is both less than 2 and greater than 5. In the examples we study, however, every algebraic sentence will describe one and only one geometric figure.

The *graph* of an algebraic sentence about x and y is the set of all points in the plane whose coordinates make the sentence a true statement. For example, consider the statement $x > 0$. The point $(2, 1)$ is in the graph of $x > 0$, since $2 > 0$ is a true statement. The point $(0, 4)$ is *not* in the graph of $x > 0$, since $0 > 0$ is *not* a true statement. The graph of this sentence is the set of all points whose x-coordinate is positive. This would be all of the points in the plane to the right of the y-axis, a half-plane. (Figure 11-4)

The graph of the sentence $y = 2$ is the set of all points whose y-coordinate is equal to 2. The x-coordinate can be any number, since there are no restrictions on x. This is a line parallel to and 2 units above the x-axis. (Figure 11-5)

From these examples we see that a graph is a geometric figure, a set of points in the plane. In learning about graphs we will be interested in two problems:

(1) Given an algebraic sentence, to draw its graph.
(2) Given a geometric figure, to find an algebraic sentence for which it is the graph.

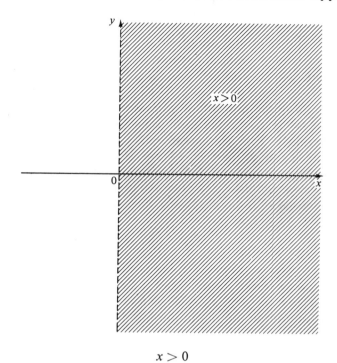

$$x > 0$$

Figure 11-4

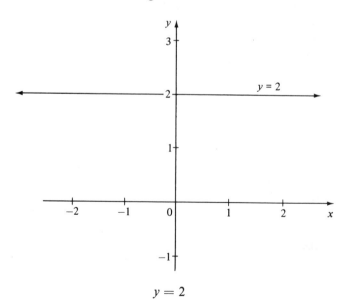

$$y = 2$$

Figure 11-5

4. Graphs of Lines, Rays, and Line Segments

An algebraic sentence of the form $Ax + By + C = 0$, where A, B, and C are real numbers and A and B are not both zero is called a *linear equation* in x and y. The graph of a linear equation is a *line* in the plane, and conversely, any line in the plane is the graph of some linear equation. If $A = 0$ the equation takes the form $y = k$, where k is a constant, and its graph is a line parallel to the x-axis passing through the point $(0, k)$. If $B = 0$ the equation is of the form $x = h$, where h is a constant, and its graph is a line parallel to the y-axis passing through the point $(h, 0)$. (Figure 11-6)

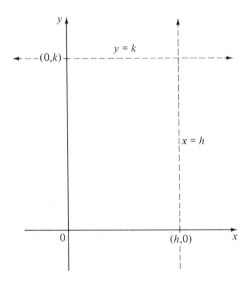

Graphing the lines $x = h$ and $y = k$

Figure 11-6

Since two points determine a line, to draw the graph of a linear equation we need to find only two points whose coordinates will make the equation a true statement. We often say such points "satisfy" the equation.

For example, consider the linear equation $x + y = 1$. The point $(1, 0)$ will satisfy this equation since $1 + 0 = 1$ is a true statement. The point $(0, 1)$ also satisfies the equation. The graph then is a line passing through the two points $(0, 1)$ and $(1, 0)$. (Figure 11-7)

Because of the one-to-one correspondence between points in the plane and ordered pairs of numbers, we can describe the graph in two ways:

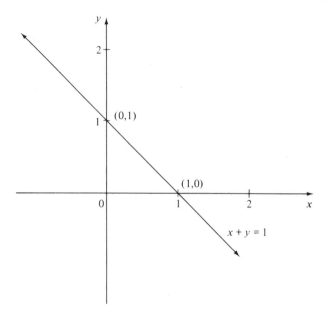

The line $x + y = 1$

Figure 11-7

by a geometric figure, or in the terms of set theory, as the set $\{(x, y) \mid x + y = 1\}$. Just as we speak of "the point (x, y)," we will often refer to "the line $Ax + By + C = 0$."

To find an algebraic sentence whose graph is a ray or a line segment, we can make use of inequalities. For example, consider the ray pictured in Figure 11-8.

This ray is parallel to and one unit above the x-axis, hence the y-coordinate of every point on the graph is 1. However, the x-coordinate is restricted. Only points with non-negative abscissas lie on the graph. To describe the ray we need two algebraic sentences

$$y = 1 \quad \text{and} \quad x \geqq 0$$

The two sentences are connected by the word *and*, since to be in the graph, a point must satisfy *both* conditions; i.e., its ordinate must be 1 and its abscissa greater than or equal to zero. This graph is the set $\{(x, y) \mid y = 1 \text{ and } x \geqq 0\}$.

In the same way, the line segment whose end-points are $(0, 1)$ and $(1, 1)$ can be described by the sentences

$$y = 1 \quad \text{and} \quad 0 \leqq x \leqq 1$$

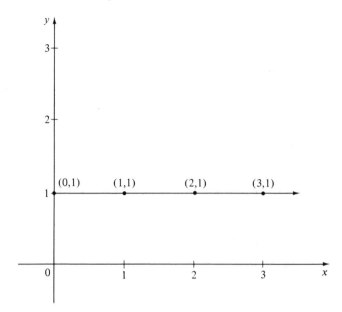

The set $\{(x, y) \mid y = 1 \text{ and } x \geqq 0\}$

Figure 11-8

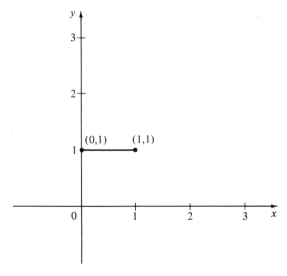

The set $\{(x, y) \mid y = 1 \text{ and } 0 \leqq x \leqq 1\}$

Figure 11-9

Exercise 11.4

1. Draw the graph of $Ax + By + C = 0$, if
   (a) $A = 0; B = 2; C = 4$ (b) $A = 3; B = 0; C = -7$
   (c) $A = 1; B = 2; C = -1$ (d) $A = -2; B = 3; C = 0$
   (e) $A = 0; B = 1; C = 0$

Draw the graph of each of the following algebraic sentences.

2. $x = 0$ and $y \geq 0$
3. $x = 1$ and $y \leq 1$
4. $x = -2$ and $-1 \leq y \leq 1$
5. $y = 1$ and $-4 \leq x \leq 4$

Find an algebraic sentence for each of the following graphs.

6.

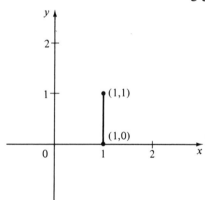

7.

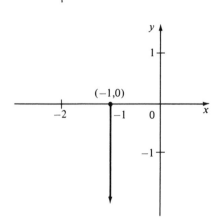

8.

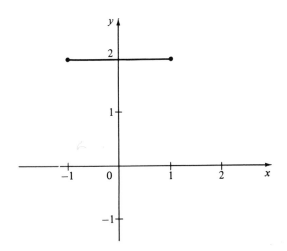

9.

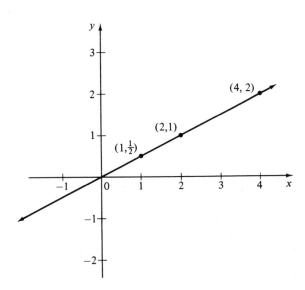

10.

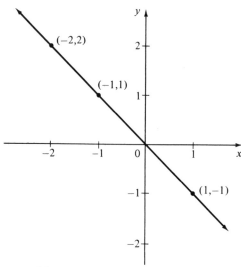

11.

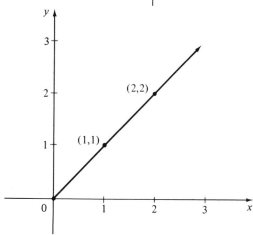

12. What is the graph of $Ax + By + C = 0$ if $A = B = 0$, but $C \neq 0$?

13. What is the graph of $Ax + By + C = 0$ if $A = B = C = 0$?

14. Draw the graph of the equations $y = x$; $y = x + 1$; $y = x - 1$ on the same coordinate axes. What do you observe about these lines? What do you surmise about the graphs of equations of the form $y = x + b$, where b is any real number?

15. Draw the graphs of the equations $y = x$ and $y = -x$ on the same axes; of $y = x + 1$ and $y = -x + 1$ on the same axes. What do you observe about these pairs of lines? Make a conjecture about the relationship between a line whose equation is of the form $y = x + a$ and a line whose equation is of the form $y = -x + b$, where a and b can be any real numbers.

16. Draw the graph of $x + y = 1$ on the coordinate axes described in Exercise 11.2, problem 11, using *parallelogram coordinates*. Draw the graph of $x + y = 1$ on the usual perpendicular coordinate axes, using rectangular coordinates. Do the same thing for $x = 2$. What difference do you notice?

5. Graphs of Other Sets in the Plane

In the last section we found algebraic sentences describing lines, rays and line segments. Other plane figures, such as angles, rectangles, and triangles can also be described in this way.

Suppose we wish to describe the set of points to the right of the y-axis and above the x-axis. (This is called the first *quadrant*.)

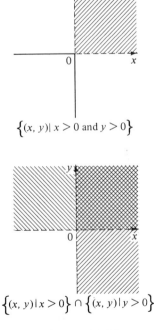

$$\{(x, y)\mid x > 0 \text{ and } y > 0\}$$

$$\{(x, y)\mid x > 0\} \cap \{(x, y)\mid y > 0\}$$

Figure 11-10

Since for all points in this region both the x-coordinate and y-coordinate are positive, this region is the set

$$\{(x, y)\mid x > 0 \text{ and } y > 0\}$$

The word "and" in the defining sentence suggests that we might also describe this set as the intersection of two sets,

$$\{(x, y)\,|\,x > 0\} \cap \{(x, y)\,|\,y > 0\}$$

The first set is the half-plane to the right of the y-axis and the second is the half-plane above the x-axis. It is not hard to see that the intersection of these two sets (the points common to both) is the first quadrant.

In drawing the graph of a region in the plane it is useful to indicate whether or not the boundary is included in the set. The usual practice is to draw the boundary with solid lines if it is included and with dotted lines if it is not.

Now let us attempt to describe the set of points in the rectangular region having vertices $(0, 0)$, $(1, 0)$, $(1, 2)$ and $(0, 2)$. (Figure 11-11)

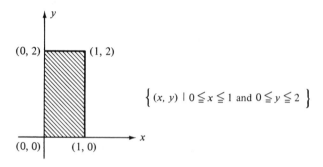

$$\left\{(x, y)\,|\,0 \leqq x \leqq 1 \text{ and } 0 \leqq y \leqq 2\right\}$$

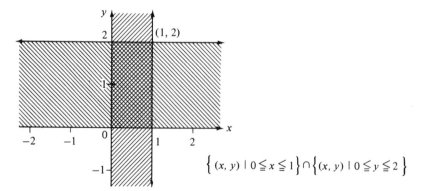

$$\left\{(x, y)\,|\,0 \leqq x \leqq 1\right\} \cap \left\{(x, y)\,|\,0 \leqq y \leqq 2\right\}$$

Figure 11-11

For all points in this set, the x-coordinate is either 0 or 1 or is between 0 and 1, while the y-coordinate is 0, 2 or is between these two numbers. Accordingly, the set is

$$\{(x, y)\,|\,0 \leqq x \leqq 1 \text{ and } 0 \leqq y \leqq 2\}$$

Again the word "and" suggests that this set may be written as the intersection of the two sets

$$\{(x, y)\,|\,0 \leq x \leq 1\} \qquad \text{and} \qquad \{(x, y)\,|\,0 \leq y \leq 2\}$$

Each of these sets is a "strip" in the plane and their intersection is the rectangular region.

The rectangle itself is, of course, a different set from the one described above. It is the *union* of four line segments whose end-points are $(0, 0)$, $(1, 0)$, $(1, 2)$, and $(0, 2)$.

$$\{(x, y)\,|\,y = 0 \text{ and } 0 \leq x \leq 1\} \cup \{(x, y)\,|\,x = 1 \text{ and } 0 \leq y \leq 2\} \cup$$
$$\{(x, y)\,|\,y = 2 \text{ and } 0 \leq x \leq 1\} \cup \{(x, y)\,|\,x = 0 \text{ and } 0 \leq y \leq 2\}$$

A triangle is also the union of line segments. For example, the triangle with vertices $(0, 0)$, $(1, 0)$, and $(1, 1)$ in Figure 11-12 is the set

$$\{(x, y)\,|\,y = 0 \text{ and } 0 \leq x \leq 1\} \cup \{(x, y)\,|\,x = 1 \text{ and } 0 \leq y \leq 1\} \cup$$
$$\{(x, y)\,|\,x = y \text{ and } 0 \leq x \leq 1\}$$

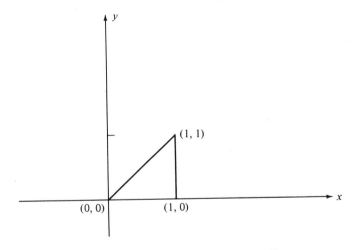

The set $\{(x, y)\,|\,y = 0 \text{ and } 0 \leq x \leq 1\} \cup$
$\{(x, y)\,|\,x = 1 \text{ and } 0 \leq y \leq 1\} \cup$
$\{(x, y)\,|\,x = y \text{ and } 0 \leq x \leq 1\}$

Figure 11-12

An angle is the union of two rays having a common end-point. The set

$$\{(x, y)\,|\,y = 0 \text{ and } x \geq 0\} \cup \{(x, y)\,|\,x = y \text{ and } x \geq 0\}$$

for example, describes the angle in Figure 11-13 whose vertex is at the origin.

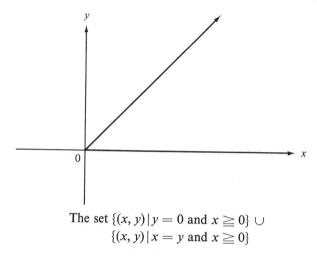

The set $\{(x, y) \mid y = 0 \text{ and } x \geqq 0\} \cup$
$\{(x, y) \mid x = y \text{ and } x \geqq 0\}$

Figure 11-13

To describe the interior of a triangle, we recall that any line separates a plane into three disjoint sets—the line itself and two half-planes on either side of the line. The line $x = 1$, for example, divides the plane into three sets—the line itself, which is the set $\{(x, y) \mid x = 1\}$ and two half-planes

$$\Pi_1 : \{(x, y) \mid x > 1\} \qquad \text{and} \qquad \Pi_2 : \{(x, y) \mid x < 1\}$$

(Figure 11-14)

The line $x = y$ divides the plane into the three sets

$$\{(x, y) \mid x = y\}; \qquad \{(x, y) \mid x > y\}; \qquad \text{and} \qquad \{(x, y) \mid x < y\}$$

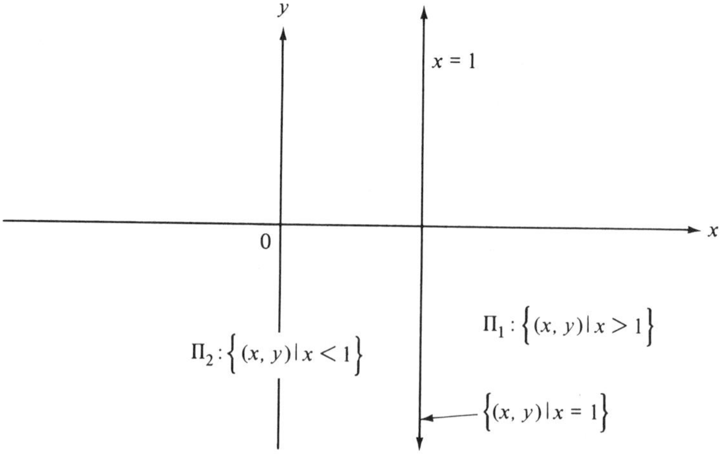

$x = 1$

$$\Pi_2 : \left\{ (x, y) \mid x < 1 \right\}$$

$$\Pi_1 : \left\{ (x, y) \mid x > 1 \right\}$$

$$\left\{ (x, y) \mid x = 1 \right\}$$

The line $x = 1$ divides the plane into three disjoint sets.

Figure 11-14

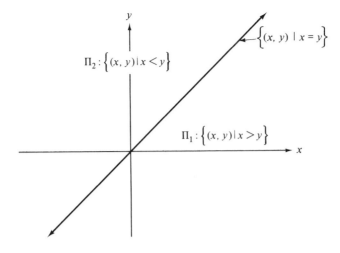

The line $x = y$ divides the plane into three disjoint sets.

Figure 11-15

The interior of the triangle with vertices $(0, 0)$, $(1, 0)$ and $(1, 1)$ can now be described as the intersection of three half-planes, one to the left of the line $x = 1$, one above the line $y = 0$, and one to the lower right of the line $x = y$. (Figure 11-16)

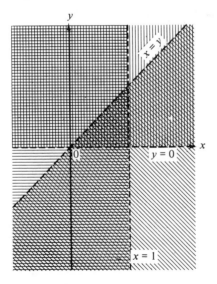

$$\{(x, y)\,|\,x < 1\} \cap \{(x, y)\,|\,y > 0\} \cap \{(x, y)\,|\,x > y\}$$

Figure 11-16

Exercise 11.5

Sketch the following graphs. Indicate in each case whether or not the boundary is included by using solid or dotted lines.

1. $\{(x, y)\,|\,x > 2\}$
2. $\{(x, y)\,|\,x > 0 \text{ and } y < 0\}$
3. $\{(x, y)\,|\,x \leq 0 \text{ and } y \leq 0\}$
4. $\{(x, y)\,|\,-1 < x < 1\}$
5. $\{(x, y)\,|\,-2 < y < -1\}$
6. $\{(x, y)\,|\,-1 < x < 0 \text{ and } 0 < y < 1\}$
7. $\{(x, y)\,|\,x \geq 0 \text{ and } x \geq y\}$
8. $\{(x, y)\,|\,x \geq 0,\ y \leq 1 \text{ and } x < y\}$

Sketch the following, then describe the graphs in terms of sets.

9. The angle with vertex at the origin containing the points $(1, 0)$ and $(1, 1)$.

10. The interior of the angle of problem 9.

11. A square measuring one unit on a side, two of whose vertices are $(0, 0)$ and $(1, 1)$.

12. The interior of the square of problem 11.

13. A square measuring one unit on a side, with center at the origin and sides parallel to the axes.

14. The interior of the square of problem 13.

15. The rectangle having vertices $(0, -1)$, $(0, -2)$, $(5, -1)$, $(5, -2)$.

16. The interior of the triangle having vertices $(0, 0)$, $(2, 0)$, $(2, 2)$.

17. The triangle having vertices $(0, 0)$, $(1, 0)$, $(0, 1)$.

18. The interior of the triangle of problem 17.

19. The line whose equation is $x = 2y$ separates the plane into three disjoint sets. The line, $\{(x, y)\,|\,x = 2y\}$, and the two half-planes $\{(x, y)\,|\,x > 2y\}$ and $\{(x, y)\,|\,x < 2y\}$. Graph the line and label each of the half-planes.

20. The line whose equation is $x + y = 2$ separates the plane into three disjoint sets. Describe each of these sets in set notation, sketch and label.

6. Slope

The idea of a slope is a common one. We speak of the slope or pitch of a roof and say that the slope of a church steeple is greater than the slope of a garage roof. The slope of a line is a measure of its "steepness." Coordinate geometry gives us a precise mathematical way of describing the slope of a line. First let us look at the graphs of the equations $y = x$, $y = 2x$, and $y = 3x$. (Figure 11-17)

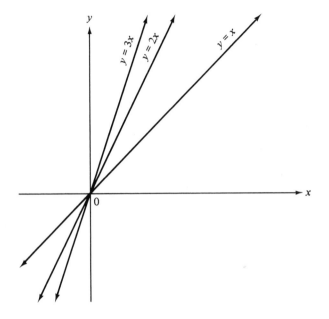

The slope of the line $y = 3x$ is greater than the slope of the line $y = 2x$, which in turn is greater than the slope of the line $y = x$.

Figure 11-17

It seems reasonable to say that the slope of the line $y = 3x$ is greater than the slope of the line $y = 2x$, which in turn is greater than the slope of the line $y = x$.

To find a numerical value for the slope of these lines, let us choose a point on the line $y = x$, say $(1, 1)$.

If we construct a perpendicular from this point to the x-axis we form a right triangle whose vertices are $(0, 0)$, $(1, 0)$ and $(1, 1)$. (Figure 11-18) The length of the vertical side of this triangle is 1 unit and the length of the horizontal side is 1 unit. The ratio of the length of the vertical side to the length of the horizontal side is $\frac{1}{1} = 1$.

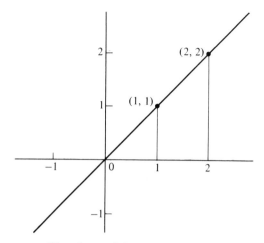

The slope of the line $y = x$ is 1.

Figure 11-18

Now choose another point, say $(2, 2)$ on the line $x = y$ and look at the right triangle whose vertices are $(0, 0)$, $(2, 0)$ and $(2, 2)$. The ratio of the length of the vertical side to the length of the horizontal side is $\frac{2}{2} = 1$. The student should convince himself that this ratio will be the same no matter what point on the line we choose.

Now let us look at the "steeper" line $y = 2x$. If we choose a point on this line, say $(1, 2)$, then the length of the vertical side of the triangle formed is 2 units and the length of the horizontal side is 1 unit. The ratio of the length of the vertical side to the length of the horizontal side is $\frac{2}{1} = 2$. This ratio will be the same for any point we choose on the line.

Since this ratio is constant for a given line and since the steeper the line the larger the ratio, this suggests that this ratio would be a good way to measure slope. We will then define the *slope of a line* to be the *ratio of the change in the y-direction* (or *vertical direction*) *to the change in the x-direction* (or *horizontal direction*).

To find this ratio, we choose any two points on the line, $p_1(x_1, y_1)$ and $p_2(x_2, y_2)$. (Figure 11-19)

The change in the y-direction is $y_2 - y_1$, while the change in the x-direction is $x_2 - x_1$. The slope of the line, then, is

$$m = \frac{y_2 - y_1}{x_2 - x_1}$$

Note that the order in which we choose the points doesn't matter since

$$\frac{y_1 - y_2}{x_1 - x_2} = \frac{y_2 - y_1}{x_2 - x_1} = m$$

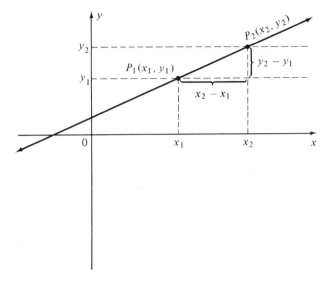

The slope of the line is $m = \dfrac{y_2 - y_1}{x_2 - x_1}$

Figure 11-19

The slope of the line $y = x$ is 1, since two points on this line are p_1 (0, 0) and p_2(1, 1) and $m = (1 - 0)/(1 - 0) = 1$. The slope of the line $y = 2x$ is 2, since p_1 (0, 0) and p_2 (1, 2) are points on this line and $m = (2 - 0)/(1 - 0) = 2$. What is the slope of the line $y = 3x$?

This definition of slope tells us several things. If a line is parallel to the x-axis, then the y-coordinate is the same for every point on the line. Thus $y_2 - y_1 = 0$, and the slope of the line is zero. On the other hand, if the line is parallel to the y-axis, the x-coordinate is the same for all points on the line, $x_2 - x_1 = 0$, and the slope is undefined, since division by zero is undefined. Thus slope is defined only for non-vertical lines.

It is clear from the above definition that slope could be negative. What geometric meaning can we give to a negative slope? Consider the line $y + x = 1$. Two points on this line are (1, 0) and $(\frac{1}{2}, \frac{1}{2})$. Its slope is $m = (\frac{1}{2} - 0)/(\frac{1}{2} - 1) = -1$.

The graph in Figure 11-20 suggests an answer to our question. If a line rises as we move from left to right, its slope is positive; if it falls, its slope is negative; if it is horizontal, its slope is zero.

If two lines are parallel, then they have the same slope; conversely, if two lines have the same slope, then they must be parallel. Thus the slope can tell us if two lines are parallel. It can also tell us if two lines are perpendicular. It is easy to see that if a line has positive slope, then any line perpendicular to it must have negative slope. In fact, if $L_1 \perp L_2$ and L_1 has slope m, then L_2 has slope $-(1/m)$. That is, if two lines are perpen-

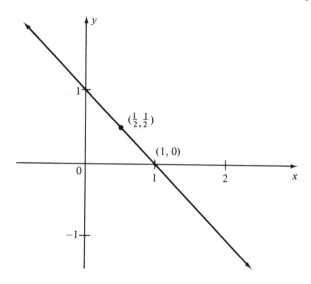

The line $x + y = 1$ has negative slope.

Figure 11-20

dicular, (and neither is a vertical line) then their slopes are negative reciprocals of each other. Conversely, if two lines have slopes which are negative reciprocals, then they are perpendicular.

For example, the lines

$$2x + y + 1 = 0$$

and

$$4x + 2y + 1 = 0$$

are parallel, since they both have slope -2. (Verify this!)

The lines

$$x - 3y + 2 = 0$$

and

$$3x + y - 1 = 0$$

are perpendicular, since their slopes are $\frac{1}{3}$ and -3, respectively. (Verify!)

We can use the slope to tell if three points are collinear. If we have three points, p, q, and r, and if the slope of the line $\overleftrightarrow{pq}$ is the same as the slope of $\overleftrightarrow{qr}$, then $\overleftrightarrow{pq} \parallel \overleftrightarrow{qr}$. Since both lines contain q, $\overleftrightarrow{pq} = \overleftrightarrow{qr}$, and p, q, and r are on one line.

For example, p (0, 3), q (2, 4), and r (−2, 2) are collinear since the slope of $\overleftrightarrow{pq}$ is $(4 - 3)/(2 - 0) = \frac{1}{2}$, and the slope of $\overleftrightarrow{qr}$ is $(2 - 4)/(-2 - 2) = \frac{1}{2}$.

We can find the slope of a line by another method. If the equation of the line can be put into the form $y = mx + b$, then the number m is the slope of the line. In other words, if we can solve for y, then the coefficient of x gives us the slope.

To see this, let (x_1, y_1) and (x_2, y_2) be two points on the line $y = mx + b$. Since these two points are on the line, they satisfy the equation and $y_1 = mx_1 + b$; $y_2 = mx_2 + b$. The slope of the line is then

$$\frac{y_2 - y_1}{x_2 - x_1} = \frac{(mx_2 + b) - (mx_1 + b)}{x_2 - x_1}$$

$$= \frac{mx_2 - mx_1}{x_2 - x_1} = \frac{m(x_2 - x_1)}{x_2 - x_1}$$

$$= m$$

In the equation $y = mx + b$, if $x = 0$, then $y = b$, that is, $(0, b)$ is a point on the line. Since this point is on the y-axis, b is called the y-intercept. We call $y = mx + b$ the slope, y-intercept form of the line and it is perhaps the most useful form for the equation of a line.

As an example, consider the line $x - 2y - 2 = 0$. Solving for y, we get $y = \frac{1}{2}x - 1$. The slope is then $\frac{1}{2}$, and the y-intercept is -1. In the equation $x = 2$, we cannot solve for y. This is a vertical line and its slope is undefined.

Exercise 11.6

1. Find the missing coordinate so that the line determined by the two points will be horizontal.

   (a) $(2, 3)$ and $(-2, \underline{\quad})$ (b) $(-3, -1)$ and $(3, \underline{\quad})$
   (c) $(8, -4)$ and $(0, \underline{\quad})$ (d) (a, b) and $(c, \underline{\quad})$

2. Find the missing coordinate so that the line determined by the two points will be vertical.

   (a) $(\underline{\quad}, 3)$ and $(-1, 2)$ (b) $(0, -4)$ and $(\underline{\quad}, 5)$
   (c) $(2, a)$ and $(\underline{\quad}, b)$ (d) (a, b) and $(\underline{\quad}, c)$

3. The grade of a road is defined to be the "rise" (the change in the vertical direction) divided by the "run" (the change in the horizontal direction). If a road rises 30 feet vertically over a run of 600 feet, what is the grade of the road?

4. The slope of a line segment is the slope of the line containing it. Find the slope of the line segments whose end-points are:

   (a) $(3, 1)$ and $(0, 0)$ (b) $(2, 5)$ and $(4, -6)$
   (c) $(0, 0)$ and $(1, -4)$ (d) $(\frac{2}{3}, \frac{1}{2})$ and $(\frac{5}{6}, \frac{3}{8})$

5. Find the slope of the lines:
   (a) $y = 5x$ (b) $y + 2x = 0$
   (c) $y - 5x + 1 = 0$ (d) $2y - 3x = 0$

6. A line has slope 1, and contains the point $(2, 5)$. If the point $(6, \underline{\hspace{1cm}})$ is on the line, find the missing coordinate. Find at least one other point on the line.

7. A line has slope $-\frac{1}{3}$ and contains the point $(1, 2)$. If the point $(0, \underline{\hspace{1cm}})$ is on the line, find the missing coordinate. Find one other point on the line.

8. A triangle has vertices $(0, 1)$, $(2, 3)$, $(-3, 2)$. Find the slope of each side. Is this a right triangle?

9. Show that the triangle with vertices $(-3, 0)$, $(3, -6)$ and $(5, 8)$ is a right triangle.

10. The vertices of a quadrilateral are $p_1(0, 0)$, $p_2(8, 0)$, $p_3(10, 4)$ and $p_4(2, 4)$. Find the slope of each side. Is this a parallelogram? Find the slope of the diagonals.

11. The vertices of a quadrilateral are $p_1(-2, 2)$, $p_2(2, -2)$, $p_3(4, 2)$ and $p_4(2, 4)$. Show that the diagonals of the quadrilateral are perpendicular.

12. The vertices of a quadrilateral are $p_1(-5, 4)$, $p_2(3, 5)$, $p_3(7, -2)$ and $p_4(-1, -3)$. Show that the quadrilateral is a parallelogram with perpendicular diagonals.

13. Test the following sets of points for collinearity.
    (a) $(1, 1)$, $(2, 2)$ and $(-5, -5)$
    (b) $(-1, 1)$, $(-3, -3)$ and $(5, 13)$
    (c) $(2, 4)$, $(6, -2)$ and $(306, -197)$
    (d) $(0, -5)$, $(2, 3)$ and $(-5, -31)$

14. Find the missing coordinate so that the three points shall be collinear.
    (a) $(2, 0)$, $(7, 1)$ and $(-3, \underline{\hspace{1cm}})$
    (b) $(0, \frac{3}{2})$, $(5, 4)$ and $(\underline{\hspace{1cm}}, -\frac{1}{2})$

15. Find the slope and the y-intercept of the following lines by putting them into the form $y = mx + b$.
    (a) $x - 2y + 7 = 0$ (b) $2x + 3y = 0$
    (c) $3x + y + 1 = 0$ (d) $y + 4 = 0$
    (e) $Ax + By + C = 0; B \neq 0$

7. Intersection of Two Lines

If two lines in a plane are not parallel, then they intersect in a single point. If we know the equations of these two lines, we can find the coordinates of the point of intersection by algebraic methods.

For example, to find the point of intersection of the two lines

$$x + y - 5 = 0$$

and $$3x - 2y = 0$$

(Figure 11-21) we want a point (x, y) which lies on both lines; i.e., a point (x, y) which satisfies both equations. Suppose we multiply the first equation by 2. We now have the two equations

$$2x + 2y - 10 = 0$$
$$3x - 2y = 0$$

Now if a point satisfies *both* of these equations, it satisfies their sum. If we add these two equations together we get

$$5x - 10 = 0 \quad \text{or} \quad x = 2$$

This means that if (x, y) is the point of intersection of these two lines, then x must be 2. But if we know the x-coordinate of the point we can find the y-coordinate, since (x, y) satisfies both equations. Substituting $x = 2$ in the equation $x + y - 5 = 0$, we get $y = 3$. (We could just as well have substituted $x = 2$ in the equation $3x - 2y = 0$. We would still get $y = 3$.)

The point of intersection then is $(2, 3)$. The student should check to see that this point does in fact lie on both lines by showing that it satisfies both equations.

The above method of solution raises an interesting point. We multiplied the first equation by 2. Does this change the graph of the line in any way? To put it another way, are the two sets

$$\{(x, y) \mid x + y - 5 = 0\} \quad \text{and} \quad \{(x, y) \mid 2x + 2y - 10 = 0\}$$

the same?

It is not hard to see that any point in the first set is also in the second and conversely. If we write the second equation as $2(x + y - 5) = 0$, we can see that a point will satisfy this equation if and only if it satisfies the equation $x + y - 5 = 0$. Thus $x + y - 5 = 0$ and $2x + 2y - 10 = 0$ are equations of the same line.

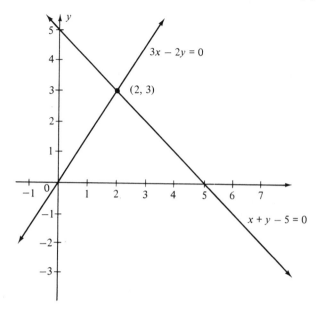

The two lines $3x - 2y = 0$ and $x + y - 5 = 0$ intersect
in the point (2, 3)

Figure 11-21

Exercise 11.7

In problems 1 through 4 find the point of intersection of the pairs of lines.
Check your answer by substitution.

1. $x + y - 5 = 0$
    $x - y = 0$

2. $x + y - 5 = 0$
    $x + 2y = 0$

3. $2x - y + 1 = 0$
    $x + 3y + 4 = 0$

4. $x - 2y + 1 = 0$
    $2x + 3y + 7 = 0$

5. Check the solutions of problems 1 through 4 by graphing each pair
    of lines.

6. In trying to find the point of intersection of two lines $x + y - 4 = 0$
    and $-x - y + 1 = 0$ we arrive at the equation $-3 = 0$, which is,

of course, false. What does this indicate about the point of intersection of these two lines? Graph.

7. In trying to find the point of intersection of the two lines $x - 2y + 1 = 0$ and $-2x + 4y - 2 = 0$ we arrive at the equation $0 = 0$. What does this indicate about the point of intersection? Graph.

8. The Distance Formula

The shortest distance between two points in the plane is the length of the line segment joining them. To find this distance we use the Pythagorean Theorem.

Consider, for example, the points p (2, 1) and q (3, 4) in Figure 11-22.

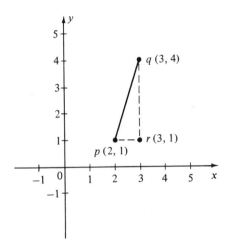

The distance from p to q is found by the Pythagorean Theorem.

Figure 11-22

First we construct a right triangle prq by drawing $\overline{qr}$ and $\overline{pr}$ parallel to the y and x-axes, respectively. The point r has the same x-coordinate as q and the same y-coordinate as p. It is easy to see that $pr = 3 - 2 = 1$ and $qr = 4 - 1 = 3$. Since prq is a right triangle

$$(pq)^2 = (pr)^2 + (qr)^2$$
$$(pq)^2 = 1^2 + 3^2$$
$$pq = \sqrt{10}$$

Now let us develop a formula for the distance between any two points in the plane p (x_1, y_1) and q (x_2, y_2).

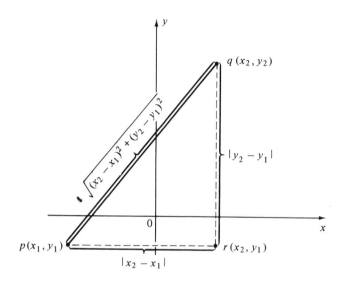

The Distance Formula

Figure 11-23

Again we construct the right triangle prq in Figure 11-23, where r has the coordinates (x_2, y_1). The distance pr is $|x_2 - x_1|$, and $qr = |y_2 - y_1|$. Distance cannot be a negative number, and since, in general, we don't know whether x_1 or x_2 is the larger, we use the absolute value bars to insure that pr will be positive. Recall that $|x_2 - x_1| = x_2 - x_1$ if $x_2 \geq x_1$, and $|x_2 - x_1| = x_1 - x_2$ if $x_1 \geq x_2$. Thus $|x_2 - x_1|$ is always positive.

By the Pythagorean Theorem

$$(pq)^2 = (pr)^2 + (qr)^2$$
$$= |x_2 - x_1|^2 + |y_2 - y_1|^2$$
$$= (x_2 - x_1)^2 + (y_2 - y_1)^2$$

and
$$pq = \sqrt{(x_2 - x_1)^2 + (y_2 - y_1)^2}$$

Note that the order in which we choose the points is unimportant, since $(x_2 - x_1)^2 = (x_1 - x_2)^2$ and $(y_2 - y_1)^2 = (y_1 - y_2)^2$.

Using the distance formula, we find that the distance between the two points $(2, 4)$ and $(1, -1)$ is

$$\sqrt{(2-1)^2 + (4-(-1))^2}$$
$$= \sqrt{1^2 + 5^2}$$
$$= \sqrt{26}$$

Since a circle is defined to be the set of all points which are a fixed distance r from a given point p, we can use the distance formula to get an algebraic sentence whose graph is a circle. Suppose we wish to describe the circle with center at the origin and radius 1 unit. If (x, y) is any point on the circle, the distance from (x, y) to the origin $(0, 0)$ must be 1, or using the distance formula

$$\sqrt{(x-0)^2 + (y-0)^2} = 1$$

Squaring both sides, we have $x^2 + y^2 = 1$, an algebraic sentence corresponding to the circle.

In general, if we wish to describe the graph of a circle with center at the point (x_0, y_0) and radius r, we want the points (x, y) whose distance from (x_0, y_0) is r:

$$\sqrt{(x-x_0)^2 + (y-y_0)^2} = r$$

or
$$(x-x_0)^2 + (y-y_0)^2 = r^2$$

We will call this the equation of the circle.

We can also use the distance formula to describe the interior of this circle. Since the distance from (x_0, y_0) to points in the interior of the circle must be *less than* r, the interior of the circle is the set

$$\{(x, y) \mid (x-x_0)^2 + (y-y_0)^2 < r^2\}$$

and the exterior of the circle is the set

$$\{(x, y) \mid (x-x_0)^2 + (y-y_0)^2 > r^2\}$$

For example, in Figure 11-24 the equation of the circle with center $(-1, 2)$ and radius 2 is $(x-(-1))^2 + (y-2)^2 = 2^2$ or $(x+1)^2 + (y-2)^2 = 4$. Its interior is the set

$$\{(x, y) \mid (x+1)^2 + (y-2)^2 < 4\}$$

and its exterior is the set

$$\{(x, y) \mid (x+1)^2 + (y-2)^2 > 4\}$$

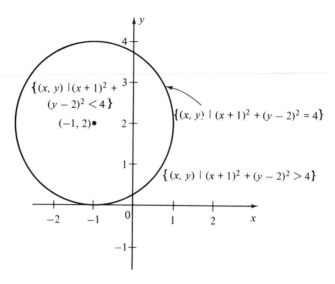

Describing a circle, its interior and its exterior by using the distance formula

Figure 11-24

Exercise 11.8

1. Find the distance between the following pairs of points.
   (a) $(0, 0)$ and $(3, 4)$ (b) $(1, 2)$ and $(1, 14)$
   (c) $(3, 8)$ and $(-5, -7)$ (d) $(-2, 3)$ and $(-1, 4)$
   (e) $(10, 1)$ and $(-1, -1)$ (f) $(-6, 3)$ and $(4, -2)$

2. Show that the triangle having vertices $(-1, 0)$, $(7, -8)$ and $(-2, -9)$ is isosceles.

3. Show that the triangle having vertices $(0, 0)$, $(3, 4)$ and $(-1, 1)$ is isosceles.

4. The vertices of a quadrilateral are $p_1(-1, -1)$, $p_2(1, 0)$, $p_3(2, 2)$, and $p_4(0, 1)$. Show that the quadrilateral is a rhombus.

5. Use the distance formula to show that the triangle with vertices $(-3, 0)$, $(3, -6)$ and $(5, 8)$ is a right triangle.

6. Use two methods to show that the triangle with vertices $(1, 1)$, $(3, 0)$ and $(4, 7)$ is a right triangle.

7. Use the distance formula to show that the points $(-1, 6)$, $(1, 4)$ and $(7, -2)$ are collinear.

8. Use two methods to show that the points $(-1, 0)$, $(1, -1)$ and $(3, -2)$ are collinear.

9. Show that if x is any real number, then $|x|^2 = x^2$. (*Hint:* By definition $|x| = x$ if $x \geq 0$ and $|x| = -x$ if $x < 0$.)

10. Find the equation of the circle
    (a) with center $(3, 2)$ and radius 4
    (b) with center $(2, -5)$ and radius 5
    (c) with center $(-5, -2)$ and radius 6
    (d) with center $(\frac{3}{4}, -\frac{1}{2})$ and radius $\frac{2}{3}$

11. Find the equation of the circle
    (a) with center $(4, 2)$ and passing through the origin
    (b) with center at the origin and passing through the point $(4, 2)$
    (c) with center $(2, 1)$ and passing through the point $(-1, 3)$

12. Describe in terms of sets the interior and exterior of each circle of problem 10.

13. Graph the rectangle whose vertices are $(1, 1)$, $(4, 1)$, $(4, 7)$ and $(1, 7)$. Find its perimeter and its area.

14. Graph the quadrilateral whose vertices are $(3, 0)$, $(3, 4)$, $(7, 6)$ and $(7, 0)$. Compute its perimeter and its area.

9. Locating Points in Space

We have seen that we can describe the location of any point in a plane by an ordered pair of real numbers once we have chosen a coordinate system. We can extend this idea to locate points in space. Suppose we wished to describe the location of some fixed point in the room, say a corner of the desk. First we choose a corner of the room as a reference point, or origin. To describe the location of the corner of the desk to someone you might tell him first to move four feet from the corner along the front wall, then three feet from this point parallel to the side wall, then up two and one-half feet. (Figure 11-25)

It took three numbers to describe the location of the point in space. There is a one-to-one correspondence between ordered *triples* of real numbers, (x, y, z) and points in space. To set up a coordinate system in space we need three number lines, each perpendicular to the other two.

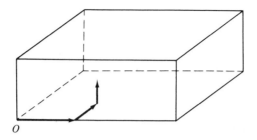

Locating a point in the room by giving three numbers

Figure 11-25

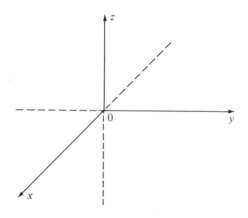

Coordinate axes in space

Figure 11-26

We will call these the x, y and z-axes, respectively. The point of intersection is called the origin and the arrows indicate the positive directions. (Figure 11-26)

Note that on the number line, which has dimension one, it takes only one real number to locate a point. In the two-dimensional plane we can locate a point with an ordered *pair* of real numbers. In space, which has three dimensions, three real numbers are required to describe the location of a point.

To locate the point (2, 1, 3), for example, we would move 2 units in a positive direction on the x-axis, one unit in a positive direction parallel to the y-axis, then 3 units in the positive direction parallel to the z-axis. This can be pictured (Figure 11-27) as the outermost corner of a box whose dimensions are 2, 1 and 3.

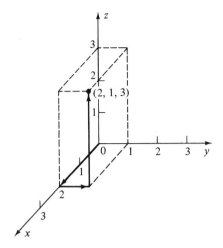

Locating the point (2, 1, 3)

Figure 11-27

We can locate the point $(-1, 3, 2)$ in the same way, only this time we move one unit in the *negative* direction along the x-axis, as indicated by the -1. (Figure 11-28)

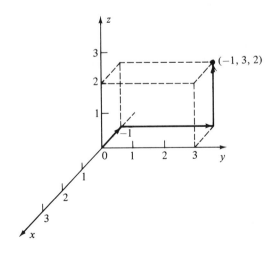

Locating the point $(-1, 3, 2)$

Figure 11-28

Exercise 11.9

In each of the following, sketch a coordinate system in space and locate the given point.

1. $(1, 2, 2)$
2. $(-1, 1, 1)$
3. $(0, -1, 2)$
4. $(-1, -1, -1)$
5. $(0, 0, 2)$

10. Locating Points on the Surface of the Earth

Points on the surface of the earth can be located by ordered pairs of numbers. Since the surface of the earth is not flat, but is approximately spherical, we cannot use rectangular coordinates. Instead, we use a network of intersecting circles.

One set of circles, called *meridians of longitude*, passes through the North and South Poles. (Figure 11-29) These are great circles (intersections of the sphere and planes through the center of the sphere).

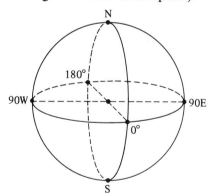

Meridians of Longitude

Figure 11-29

These great circles are assigned numbers 0 through 180. The meridian of longitude passing through Greenwich, England, is called the *prime meridian* and is labeled 0°. The semi-circle on the opposite side of the globe is labeled 180°. Those meridians in between are labeled as a certain number of degrees east or west of the prime meridian. Looking down on the North Pole, the meridians are numbered as indicated in Figure 11-30.

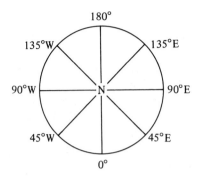

Looking down on the North Pole at the meridians of longitude

Figure 11-30

This gives us one coordinate. The second family of circles is called *parallels of latitude*. The equator is one of the parallels of latitude. Other parallels of latitude are formed by taking the intersection of the sphere with planes parallel to the plane of the equator. These are not great circles

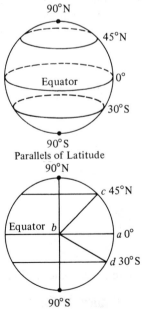

Figure 11-31

(with the exception of the equator) since these planes do not pass through the center of the sphere.

The parallels of latitude are assigned numbers from 0 to 90. The equator is labeled 0°, the North Pole 90° N and the South Pole 90° S. The other parallels of latitude are labeled as a certain number of degrees north or south of the equator.

To see how these numbers are determined, consider the cross section of the sphere in Figure 11-31. If m $\angle cba = 45$ where b is the center of the earth and a is on the equator, then the parallel of latitude passing through point c is labeled 45° N. If m $\angle abd = 30$, then the parallel of latitude passing through point d is labeled 30°S.

A point on the surface of the earth is located by giving its longitude and latitude. For example, the location of Denver, Colorado, is approximately 105° W, 40° N. In this coordinate system, the North Pole and South Pole can be described uniquely by their latitudes alone—90° N and 90° S. They need no longitude measurement to describe their locations.

11. Conic Sections

The Greeks studied the conic sections extensively. It is believed that these curves were discovered by Menaechmus (ca. 350 B.C.) for the purpose of solving the problem of the duplication of the cube. It was Apollonius (ca. 262–200 B.C.), however, who is considered the "father" of conic sec-

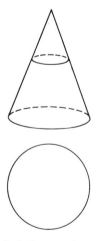

The circle is a conic section.

Figure 11-32

tions. In his eight volume work, *Conic Sections*, he investigated these curves thoroughly.

Apollonius obtained these curves by taking cross sections of a hollow right circular cone. We have already seen that a cross section parallel to the base is a circle.

If we cut the cone at an angle, as in Figure 11-33, the cross section is a curve called an *ellipse*. The larger the angle at which the cut is made, the more elongated the ellipse will be. If the cut is made nearly parallel to the base, the ellipse will look very much like a circle.

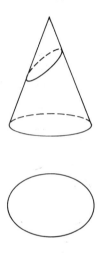

The ellipse

Figure 11-33

Now let us take a cross section parallel to the slanted edge of the cone and at the same time, let us think of the base of the cone as infinitely far away from the vertex. This cross section, which is an unbounded curve, is called a *parabola*. (Figure 11-34)

For the final conic section suppose that we have *two* cones placed vertex to vertex. This time we take the cross section so that it cuts through both cones, and again we think of both cones as extending infinitely far in both directions. The resulting curve which has two parts or branches is called a *hyperbola*. It is an unbounded curve. Although to the student one branch of the hyperbola may look very much like the parabola, the two curves are totally different. They are *not* congruent. In fact, even if you took a very small section of a parabola, it would be impossible to fit it *exactly* on any portion of the hyperbola. (Figure 11-35)

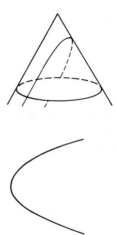

The parabola

Figure 11-34

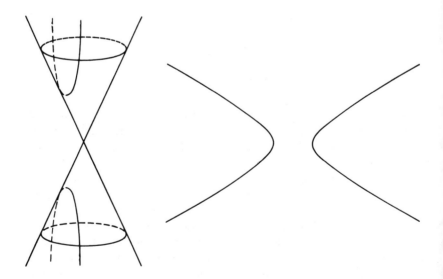

The hyperbola

Figure 11-35

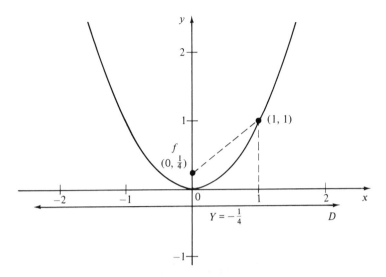

The parabola $y = x^2$

Figure 11-36

Now let us describe these same curves in a different way using the ideas of coordinate geometry.

Consider the parabola in Figure 11-36. Let D be the line $y = -\frac{1}{4}$ and f the point $(0, \frac{1}{4})$. If p is *any* point on the curve, then the distance from p to the point f is the same as the distance from p to the line D. The point f is called the *focus* of the parabola and the line D is called its *directrix*. The equation of this parabola is $y = x^2$. A parabola can be described as a set of points having the property that if p is any point on the parabola, then the distance from p to a fixed point f is the same as the distance from p to a fixed line D.

The ellipse in Figure 11-37 has the equation $x^2/25 + y^2/9 = 1$. If p is any point on the ellipse then the sum of the distances from p to $f_1(4, 0)$ and to $f_2(-4, 0)$ is constant, in this case 10. We call f_1 and f_2 the *foci* of the ellipse. An ellipse can be described as a set of points having the property that if p is any point on the ellipse, then the *sum* of the distances from p to two fixed points f_1 and f_2 is constant.

This description of the ellipse suggests a simple way of sketching the curve. Thumbtack the ends of a piece of string to a sheet of cardboard, leaving some slack. Now with a pencil held so that the string is taut (Figure 11-38) trace the curve.

If p is any point on the hyperbola of Figure 11-39 then the *difference* of the distances from p to $f_1(5, 0)$ and to $f_2(-5, 0)$ is constant, and f_1 and

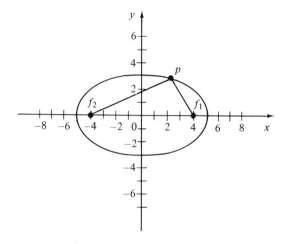

The ellipse $\dfrac{x^2}{25} + \dfrac{y^2}{9} = 1$

Figure 11-37

Tracing an ellipse using a piece of string thumbtacked to a piece of cardboard

Figure 11-38

f_2 are called the *foci* of the hyperbola. The equation of this hyperbola is

$$\frac{x^2}{16} - \frac{y^2}{9} = 1$$

A hyperbola is a set of points such that if p is any point on the hyperbola the difference of the distances from p to two fixed points f_1 and f_2 is a constant.

The conic sections have important scientific applications. If you throw a rock (not straight up) its path will be very nearly parabolic. In a vacuum its path would be precisely a parabola. Kepler showed that the planets move about the sun in elliptical orbits. If an object enters our gravitational

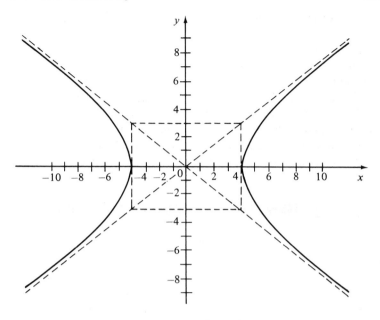

The hyperbola $\dfrac{x^2}{16} - \dfrac{y^2}{9} = 1$

Figure 11-39

field and is captured, then its path thereafter will be elliptical or circular. If it is not captured, then its path will be hyperbolic or parabolic.

The mirror of a reflecting telescope is parabolic in shape, so that a vertical cross section (see Figure 11-40) will be a portion of a parabola.

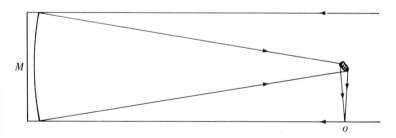

A reflecting telescope with parabolic mirror M

Figure 11-40

This parabolic mirror has the property that the parallel light rays striking it are all reflected to the same point—the focus of the parabola. A small mirror placed at this point reflects the light to a convenient position for observation.

Suppose that a room, elliptical in shape, were lined with mirrors. If you stood at one focus of the ellipse and directed the beam of a flashlight at any point on the mirrored wall, the light would be reflected so that it illuminated an object at the other focus. Sound waves behave in the same way. If you stood at one focus and whispered, a person at the other focus could hear you, although someone nearer could not.

Many natural phenomena are explained precisely in terms of the conic sections. One might even conclude that nature is partial to these curves.

New Terms Found in This Chapter

Term	Section	Term	Section
x-axis	2	y-intercept	6
y-axis	2	meridians of longitude	10
origin	2	prime meridian	10
abscissa	2	parallels of latitude	10
ordinate	2	ellipse	11
rectangular coordinates	2	parabola	11
parallelogram co-ordinates	Exercise 11.2	hyperbola	11
		focus of a parabola	11
graph	3	directrix of a parabola	11
linear equation	4	foci of an ellipse	11
quadrant	5	foci of a hyperbola	11
slope	6		

Bibliography

Courant, R., and H. Robbins. *What is Mathematics?* New York: Oxford University Press, Inc., 1941.

Eves, Howard. *An Introduction to the History of Mathematics*, rev. ed. New York: Holt, Rinehart & Winston, Inc., 1964.

Hyatt, Herman R., and Charles C. Carico. *Modern Plane Geometry for College Students*. New York: The Macmillan Company, 1967.

Kline, Morris. *Mathematics in Western Culture*. New York: Oxford University Press, Inc., 1953.

Maxwell, E. A. *Fallacies in Mathematics*. Cambridge: Cambridge University Press, 1963.

Moise, Edwin E. *Elementary Geometry from an Advanced Standpoint*. Reading, Massachusetts: Addison-Wesley Publishing Co., Inc., 1963.

Prenowitz, Walter, and Meyer Jordan. *Basic Concepts of Geometry*. Waltham, Massachusetts: Blaisdell Publishing Company, 1965.

Wolff, Peter. *Breakthroughs in Mathematics*. New York: The New American Library, Inc., 1963.

ANSWERS AND HINTS TO SELECTED PROBLEMS

Chapter 1

Exercise 1.2

1. (a) $s \in Q$; (b) $t \notin Q$; (c) $p \in Q$; (d) $r \in Q$; (e) $a \notin Q$; (f) $c \notin Q$.

3. (a) $E \cup O = N$; $E \cap O = \varnothing$
   (b) $B \cup G = \{x \mid x \text{ is a student in the freshman class}\}$
   $B \cap G = \varnothing$
   (c) $A \cup B = A$; $A \cap B = B$
   (d) $R \cup S = R$; $R \cap S = S$
   (e) $E \cup F = \{x \mid x \text{ is a multiple of 2 or } x \text{ is a multiple of 3}\}$
   $E \cap F = \{x \mid x \text{ is a multiple of 6}\}$

7. 12; 12; $n \cdot m$; $n \cdot m$

9. They are the same set.

Exercise 1.3

1. A typical equivalence class would be the set of all students in this classroom who are 18 years old.

3. A typical equivalence class would be the set of all books written by Dickens.

5. A typical equivalence class would be the set of all sets having one member.

7. R, T

9. T

11. S

13. equals

Exercise 1.4

1. No

3. (a) If two angles are right angles, then they are congruent.
   (b) If two triangles are congruent, then they are similar.
   (c) If two angles are base angles of an isosceles triangle, then they are congruent.
   (d) If three angles are the angles of some triangle, then the sum of their measure is 180°.

304

   (e) If two integers are both even, then their sum is an even integer.

5. (a) If two angles are not congruent, then they are not both right angles.

   (b) If two triangles are not similar, then they are not congruent.

   (c) If two angles are not congruent, then they are not the base angles of an isosceles triangle.

   (d) If the sum of the measures of three angles is not 180°, then they are not the angles of a triangle.

   (e) If the sum of two integers is not an even integer, then the two integers are not both even.

7. (a) If two angles are vertical angles, then they are congruent.

   (b) Two angles are vertical angles only if they are congruent.

   (c) The fact that two angles are vertical angles is a sufficient condition for them to be congruent.

   (d) The fact that two angles are congruent is a necessary condition for them to be vertical angles.

9. If beggars ride, then wishes are horses.
    If beggars do not ride, then wishes are not horses.
    If wishes are not horses, then beggars do not ride.

11. The inverse, $\sim p \Rightarrow \sim q$.

13. (a) $x = 12$

   (c) 11 cannot be written as the sum of two primes.

   (d) $x = 2, 3$

Chapter 2

Exercise 2.3

1. (a) $L_1 \cap L_2 = \varnothing$ (b) $L \subset \Pi$

   (c) $L \subset \Pi$ (d) $p \in L$

   (e) $p \in L$ (f) $p \in \Pi$

   (g) $L \cap \Pi = \{p\}$ (h) $L_1 \cap L_2 = \{p\}$

3. (a) $\overline{pq} \cap \overline{qr} = \{q\}$ (b) $\overline{pr} \cap \overline{qr} = \overline{qr}$

   (c) $\overline{pq} \cup \overline{qr} = \overline{pr}$ (d) $\overline{pq} \cup \overline{pr} = \overline{pr}$

7. There is more than one answer to each of the following. Only one answer is given.

   (a) $\overrightarrow{ad}, \overrightarrow{da}$ (b) $\overrightarrow{ad}, \overrightarrow{bd}$

   (c) $\overrightarrow{bd}, \overrightarrow{cd}$ (d) $\overrightarrow{ca}, \overrightarrow{ba}$

   (e) $\overrightarrow{cd}, \overrightarrow{ca}$ (f) $\overrightarrow{cd}, \overrightarrow{ba}$

   (g) $\overrightarrow{bd}, \overrightarrow{ca}$ (h) $\overrightarrow{bd}, \overrightarrow{ca}$

   (i) $\overrightarrow{ad}, \overrightarrow{ba}$

9. (a) $\overline{ae}$; (b) $\overline{ad}$; (c) $\{b\}$; (d) $\overline{df}$; (e) $\varnothing$; (f) $\overrightarrow{dc}$; (g) $\overline{bc}$; (h) $\overline{bc}$; (i) $\overline{ce}$

Exercise 2.4

1. yes, yes, no, yes

3. Infinitely many. Infinitely many.

5. If there are n points in a plane, no three of which are collinear, then $\frac{1}{2}n(n-1)$ lines are determined.

7. 4; $4 + 6 = 10$.

9. It is reflexive and symmetric but not transitive.

Exercise 2.5

1. Two "lines" are parallel if they lie in the same "plane" and their intersection is empty. There are no parallel lines in this model since any two lines which lie in the same plane have a non-empty intersection.

3. Theorem 1. Two distinct clubs have at most one member in common.
   Theorem 3. Given a club L and a person p not belonging to L, there is exactly one community containing both of them.
   Theorem 4. If two clubs have one or more members in common, then they both belong to exactly one community. Two clubs are "parallel" if they are in the same community and have no members in common. There are no parallel lines in this model. Persons are "noncollinear" if they do not all belong to the same club.

Exercise 2.6

1. Their intersection might be empty, a single point, a line segment or a ray.

3. Π_1; Σ_2; $\Sigma_1 \cap \Pi_2$; $L_1 \cup L_2$

5. They are either parallel or skew.

7. The intersection could be a line; the intersection could be a single point; all three planes could be parallel; two planes could be parallel and the third not; none of them parallel, yet their intersection empty.

9. It is reflexive and symmetric but not transitive.

Exercise 2.7

3. (a) T; (b) T; (c) T; (d) T; (e) F; (f) T; (g) T; (h) F; (i) F; (j) T

5. n concurrent lines form $n(n-1)$ pairs of vertical angles.

9. (a) F; (b) T; (c) F; (d) T; (e) F; (f) F; (g) T

11. A polygon with n sides has $\dfrac{(n-3)n}{2}$ distinct diagonals.

Exercise 2.9

1. (a) closed, simple   (b) not closed
   (c) not closed   (d) not closed
   (e) closed, not simple   (f) closed, not simple
   (g) closed, simple

3. (a) no; (b) yes; (c) no; (d) yes; (e) no; (f) no;
   (g) yes; (h) yes, by definition the empty set is convex

5. yes; no

7. (a) $\varnothing$; (b) $\overline{xy} \subset$ interior $\triangle abc$; (e) no

Chapter 3

Exercise 3.2

1.

p	q	pq
-1	$1\frac{1}{2}$	$2\frac{1}{2}$
0	-7	7
-10	-3	7
$2\frac{1}{2}$	14	$11\frac{1}{2}$
-2	14	16

5. $-\frac{2}{3}$; $-\frac{1}{2}$; $\frac{1}{4}$; 2; $-\frac{1}{6}$; $\frac{7}{12}$; $\frac{7}{3}$; $\frac{3}{4}$; $\frac{5}{2}$; $\frac{13}{4}$

7. Any point x on the perpendicular bisector of $\overline{pq}$ is equidistant from p and q.

9.

p	q_1	q_2	length
0	2	-2	2
2	5	-1	3
$1\frac{1}{2}$	$\frac{7}{4}$	$\frac{5}{4}$	$\frac{1}{4}$
$\sqrt{2}$	$\sqrt{2}+2$	$\sqrt{2}-2$	2
-1	$\frac{3}{2}$	$-\frac{7}{2}$	$2\frac{1}{2}$
-7	-3	-11	4

Exercise 3.3

3. A sphere

Exercise 3.4

1. 25 dexes $= 90°$; $12\frac{1}{2}$ dexes $= 45°$; $8\frac{1}{3}$ dexes $= 30°$;
   $37\frac{1}{2}$ dexes $= 135°$; 1 degree $\doteq .28$ dexes; 1 dex $= 3.6°$

3. $90°$; $45°$; $30°$; $60°$; approximately $143°$; approximately $172°$

5. (a) $45°$; (b) $15°$; (c) $60°$; (d) $85°$; (e) $10°$; (f) $40°$;
   (g) $75°$; (h) $90°$; (i) $180°$; (j) $85°$

7. $90°$; $45°$

Exercise 3.7

   1. $m\angle 2 = 90°$ because $\angle 1$ and $\angle 2$ are supplementary angles; (d) four

   5. If an acute angle has measure n, then its complement has measure $90 - n$. See Theorem 1, Section 6.

   7. They are congruent by Theorem 1 since they are both supplementary to $\angle 1$.

   9. $90°$; $\quad m\angle 1 + m\angle 2 + m\angle 3 = 180°$; $\quad \overrightarrow{ab} \perp \overrightarrow{ac}$

 11. $90°$

 13. no; yes; no

Chapter 4

Exercise 4.3

   3. There are six congruences. Two of them are $abc \leftrightarrow bac$; $abc \leftrightarrow bca$.

   5. No

   7. No

   9. No. The sum of any three of the numbers must be greater than the fourth.

 11. Yes

 13. yes; no

 15. Use the ASA Axiom.

 17. They are the same set.

 19. Use the fact that if two angles are congruent then their supplements are congruent.

 21. No

Chapter 6

Exercise 6.1

   1. 1 and 5, 2 and 6, 3 and 7, 4 and 8 are corresponding angles; 1 and 7; 2 and 8 are alternate interior angles.

   7. $\overleftrightarrow{ab}$ is not a transversal of $\overleftrightarrow{ad}$ and $\overleftrightarrow{ac}$ since it does not intersect these lines in two different points.

Exercise 6.2

   3. $\angle 1$ is supplementary to $\angle 2$ and $\angle 2 \cong \angle 3$. Therefore $\angle 1$ is supplementary to $\angle 3$.

   7. $m\angle 2 = m\angle 3 = m\angle 4 = m\angle 5 = m\angle 6 = 45°$; $\quad m\angle 7 = 135°$; $m\angle 8 = 90°$

Exercise 6.3

1. (a) 130; (b) 31; (c) $180 - 2r$; (d) 60; (e) 90; (f) 90

9. 36, 72, 72

11. (a) 80; (b) 91; (c) 100; (d) 60; (e) 4

13. $m\angle 1 = 30$; $m\angle 2 = 45$; $m\angle 3 = 80$; $m\angle 4 = 55$; $m\angle 5 = 55$;
    $m\angle 6 = 125$; $m\angle 7 = 25$; $m\angle 8 = 80$; $m\angle 9 = 75$; $m\angle 10 = 45$

15. n; $n - 2$; $180(n - 2)$

17. (a) 17; (b) 9; (c) 13; (d) 25

19. 12; 5

23. Each angle of the polygon has measure $\dfrac{(n - 2)180}{n}$.
    Since x of these angles must fill out an angle of 180°, then $x = 180/[(n - 2)180/n] = n/(n - 2)$. The number of polygons meeting at each vertex then must be $1 + n/(n - 2)$. Since this number must be a positive integer, n can be only 3 or 4.

Exercise 6.5

7. 65, 115, 115

11. (a) 25, 25, 130; (b) 65, 65, 50

13. (a) always; (b) sometimes; (c) always; (d) always;
    (e) sometimes; (f) always; (g) never; (h) never;
    (i) sometimes; (j) never

Chapter 7

Exercise 7.1

1. (a) $\frac{1}{3}$; (b) $\frac{1}{4}$; (c) 3; (d) $\frac{3}{2}$; (e) $\frac{4}{3}$; (f) $x/3y$;
   (g) $\sqrt{2}/2 = 1/\sqrt{2}$

3. $x = 2\frac{1}{2}$; $y = 6\frac{1}{2}$

5. $6\frac{2}{3}$; 12

7. $\frac{3}{2}, \frac{9}{2}, \frac{15}{2}, 9, \frac{21}{2}$

11. Use Theorem 1. Express $a, c \sim b, d$ as an equation between fractions and then subtract one from both sides of the equation.

Exercise 7.2

3. $\angle d \cong \angle a$ since they are corresponding angles. Then $\triangle abc \sim \triangle dec$ by the AA Similarity Theorem.

5. Yes. $abc \longleftrightarrow fde$

7. Yes. $abc \longleftrightarrow dfe$

11. The angles of any isosceles right triangle must measure 45, 45, 90.

13. $21\frac{3}{7}$ feet

Exercise 7.3

1. $ac = 5$; $cb = 6\frac{2}{3}$; $db = 5\frac{1}{3}$
3. $ac = 2\sqrt{5}$; $cb = 4\sqrt{5}$; $cd = 4$
7. $2\sqrt{2}$
9. $x = \frac{4}{3}$; $y = \sqrt{65}$
11. If $u = 3$, $v = 2$, then $a = 12$; $b = 5$; $c = 13$

Exercise 7.4

1. $\sin a = \cos b = \frac{4}{5}$; $\cos a = \sin b = \frac{3}{5}$; $\tan a = \frac{4}{3}$; $\tan b = \frac{3}{4}$
3. 45; $n\sqrt{2}$; $\sin a = \cos a = n/(n\sqrt{2}) = 1/\sqrt{2}$; $\tan a = n/n = 1$
5. As $\text{m}\angle a$ gets closer to 90, $\sin a$ gets closer to 1. No.
7. Small;   it gets larger;   yes.
9. $\sin a = .6$; $\cos a = .8$; $\tan a = .75$

Exercise 7.5

1. Approximately 122 feet
3. Approximately 25 feet
5. Approximately 456 feet

Chapter 8

Exercise 8.2

1. 70 sq. ft.; $13\frac{1}{8}$ sq. in.;   12 ft.; $13\frac{1}{3}$ in.
3. It is doubled;   it is multiplied by 4;   unchanged.
5. 10 square units
7. $3\frac{1}{4}$ square units
9. $a/\sqrt{2}$
11. 185
15. $3\frac{1}{8}$

Exercise 8.3

1. (a) it would be doubled;   (b) unchanged;
    (c) multiplied by 4;   (d) doubled;
    (e) multiplied by 4;   (f) unchanged
7. $3\sqrt{5}$
9. They all have area 15 square units.

11. $6 + \sqrt{3}$

17. $4\sqrt{3}$; $2\sqrt{3}$

19. Both statements are false.

21. (a) $(h_1 - h_2)(b_1 + b_2)/2$
    (b) $\frac{1}{2}(h_1 b_1) - \frac{1}{2}(h_2 b_2)$

23. $\theta = 36°$; $A = 1.718s^2$

25. $\theta = 18°$; $A = 7.70s^2$

27. 2; 4

Exercise 8.6

1. $\frac{25}{9}$; 5

3. 216; 96

Chapter 9

Exercise 9.2

5.

Base	F	E	V
triangle	4	6	4
quadrilateral	5	8	5
pentagon	6	10	6
hexagon	7	12	7
.			
.			
.			
n-gon	$n + 1$	$2n$	$n + 1$

$F + V = E + 2$

9. The set of (solid) right circular cylinders is a subset of the set of cylinders. It has no members in common with either the set of prisms or the set of solid polyhedra.

11. The area of one face is $\sqrt{3}/4$. The total surface area is $4(\sqrt{3}/4) = \sqrt{3}$ square units.

13. The area of one face is $\sqrt{3}$. The total surface area is $20\sqrt{3}$ square units.

15. $2\sqrt{3} + 3(10) = 2\sqrt{3} + 30$

Exercise 9.3

1. $F = 16$; $E = 32$; $V = 16$

Exercise 9.7

1. The volume of a half-inch cube is $(\frac{1}{2})^3 = \frac{1}{8}$ of a cubic inch.

5. It is multiplied by 8. It is multiplied by 4.

7. Volume = 64 cubic inches. Surface Area = $80 + 16\sqrt{5}$ sq. in.

9. 82,944,000 cubic feet.

11. 3 units.

13. (a) $fb = \sqrt{3}/2$;   (b) $eb = \sqrt{3}/3$;   (c) $de = \sqrt{\tfrac{2}{3}}$;
     (d) $\sqrt{3}/4$ square units;   (e) $\sqrt{2}/12$ cubic units

15. If two pyramids have equal volumes, then they have equal base areas
and equal heights. This is false.

17. (b) $\tfrac{1}{3}a^2 h$
     (e) $\tfrac{1}{3}b^2(h - k)$
     (d) $\tfrac{1}{3}a^2 h - \tfrac{1}{3}b^2 h + \tfrac{1}{3}b^2 k = \tfrac{1}{3}h(a^2 - b^2) + \tfrac{1}{3}b^2 k = \tfrac{1}{3}h(a - b)(a + b)$
         $+ \tfrac{1}{3}b^2 k$

21. By problem 16 the base and the cross section are similar figures with
constant of proportionality equal to $g/h = (h - k)/h$. The ratio of the
areas is the square of this constant.
$$\text{Area } (B_k)/a = ((h - k)/h)^2$$

23. Area $(M) = \tfrac{1}{4}a$

25. No.

Chapter 10

Exercise 10.1

1. 12.6;   15.7;   4

3. Doubled;   doubled

5. The ratio would be very close to 2π.

7. $r = \sqrt{2}$; $C = 2\sqrt{2}\,\pi$.

9. $4\pi \doteq 12.56$ units.

11. Let C_1 and r_1 stand for the circumference and radius of any sphere. If
the radius is to be increased by $6'$, then the new radius $r_2 = r_1 + 6$. The
new circumference $C_2 = 2\pi r_2 = 2\pi(r_1 + 6) = 2\pi r_1 + 2\pi \cdot 6$. Thus C_2
is larger than C_1 by $12\pi \doteq 37.68'$, and this is independent of r_1.

Exercise 10.2

3. The area of the square is $(\pi r/2)^2 = (\pi^2 r^2)/4$.
The area of the circle is πr^2. If these two are equal then $(\pi^2 r^2)/4 = \pi r^2$ and $\pi = 4$.

Exercise 10.3

1. 19.6; $\sqrt{2}$

3. Approximately 38 cubic inches

5. 5

7. $2\sqrt{2}$ inches; 8 square inches

9. 12π

11. $(\frac{25}{4})\pi$; $(\frac{25}{4})\pi$

Exercise 10.4

1.

radius of base	height	volume	lateral surface area	total surface area
2	8	32π	32π	40π
15	2	450π	60π	510π
6	6	216π	72π	144π
2	$\frac{1}{2}$	2π	2π	10π

2. doubled; doubled; unchanged

5. approximately 18.4 inches

7. doubled; multiplied by 4; doubled; multiplied by 4; multiplied by 8

Exercise 10.5

1.

radius	volume	surface area
2	$(\frac{32}{3})\pi$	16π
3	36π	36π
$\frac{1}{2}$	$(\frac{1}{6})\pi$	π
10	$1{,}333\pi$	400π

3. $(\frac{1}{3})\pi$; π; $(\frac{2}{3})\pi$

5. 14,965 gallons

7. $\pi/6$ cubic feet

Chapter 11

Exercise 11.2

3. On a line parallel to and one unit below the x-axis.

5. (a) the half plane to the left of the y-axis;
   (b) the half plane above the x-axis plus the x-axis itself;
   (c) the y-axis; (d) the x-axis; (e) the origin;
   (f) the coordinate axes; (g) a line through the origin making an angle of 45° with the x-axis;
   (h) a line through the origin passing through the point $(2, 1)$

7. Each ordinate is twice the corresponding abscissa.
   $(3, 6), (-4, -8)$, etc.

9. $(-3, 1)$

Exercise 11.4

7. $x = -1$ and $y \leq 0$

9. $x = 2y$

11. $y = x$ and $x \geq 0$

13.   All the points in the plane

15.   They are perpendicular.

Exercise 11.5

9. $\{(x, y) \mid y = 0 \text{ and } x \geq 0\} \cup \{(x, y) \mid x = y \text{ and } x \geq 0\}$

11. $\{(x, y) \mid y = 0 \text{ and } 0 \leq x \leq 1\} \cup \{(x, y) \mid x = 1 \text{ and } 0 \leq y \leq 1\} \cup$
      $\{(x, y) \mid y = 1 \text{ and } 0 \leq x \leq 1\} \cup \{(x, y) \mid x = 0 \text{ and } 0 \leq y \leq 1\}$

13. $\{(x, y) \mid x = \frac{1}{2} \text{ and } -\frac{1}{2} \leq y \leq \frac{1}{2}\} \cup \{y = \frac{1}{2} \text{ and } -\frac{1}{2} \leq x \leq \frac{1}{2}\} \cup$
      $\{(x, y) \mid x = -\frac{1}{2} \text{ and } -\frac{1}{2} \leq y \leq \frac{1}{2}\} \cup \{y = -\frac{1}{2} \text{ and } -\frac{1}{2} \leq x \leq \frac{1}{2}\}$

15. $\{(x, y) \mid x = 5 \text{ and } -2 \leq y \leq -1\} \cup \{(x, y) \mid y = -1 \text{ and } 0 \leq x \leq 5\} \cup$
      $\{(x, y) \mid x = 0 \text{ and } -2 \leq y \leq -1\} \cup \{(x, y) \mid y = -2 \text{ and } 0 \leq x \leq 5\}$

17. $\{(x, y) \mid y = 0 \text{ and } 0 \leq x \leq 1\} \cup \{(x, y) \mid x + y = 1 \text{ and } 0 \leq x \leq 1\} \cup$
      $\{(x, y) \mid x = 0 \text{ and } 0 \leq y \leq 1\}$

Exercise 11.6

1.   (a)   3;   (b) -1;   (c) -4;   (d) b

3. $\frac{1}{20}$

5.   (a)   5;   (b) -2;   (c)   5;   (d) $\frac{3}{2}$

7. $\frac{7}{3}$; $(7, 0)$

9.   The slope of one side is -1, of another $+1$, therefore these two sides are perpendicular.

11.   The slope of $\overline{p_1 p_3}$ is 0, thus $\overline{p_1 p_3}$ is horizontal. The slope of $\overline{p_2 p_4}$ is undefined, i.e., it is vertical.

13.   (a)   and   (b)   are collinear.

15.   (a) $m = \frac{1}{2}, b = \frac{7}{2}$;   (b) $m = -\frac{2}{3}, b = 0$;
      (c) $m = -3, b = -1$;   (d) $m = 0, b = -4$;
      (e) $m = -A/B, \;\; b = -C/B$

Exercise 11.7

1. $(\frac{5}{2}, \frac{5}{2})$

3. $(-1, -1)$

7.   Since $0 = 0$ is always true, this means any point on the first line is on the second line and conversely. The two equations represent the same line.

Exercise 11.8

1. (a) 5; (b) 12; (c) 17; (d) $\sqrt{2}$; (e) $5\sqrt{5}$; (f) $5\sqrt{5}$

3. The lengths of the three sides are 5, 5 and $\sqrt{2}$.

5. The lengths of the three sides are $6\sqrt{2}$, $10\sqrt{2}$ and $8\sqrt{2}$. The sum of the squares of two of these is equal to the square of the third.

7. The distances between these points are $2\sqrt{2}$, $6\sqrt{2}$ and $8\sqrt{2}$. The sum of two of these distances is equal to the third, hence the points must be collinear.

9. If $x \geq 0$, $|x| = x$ and $|x|^2 = x^2$. If $x < 0$, $|x| = -x$ and $|x|^2 = (-x)(-x) = x^2$

11. (a) $(x - 4)^2 + (y - 2)^2 = 20$
    (b) $x^2 + y^2 = 20$
    (c) $(x - 2)^2 + (y - 1)^2 = 13$

13. Perimeter $= 18$; Area $= 18$

Index